Harnessing Farms
and Forests in the
Low-Carbon Economy

Contributing Authors

Gordon R. Smith, Ecofor
Bruce A. McCarl, Texas A&M University
Changsheng Li, University of New Hampshire
Joel H. Reynolds, Statistical Solutions Consulting
Roel Hammerschlag, Institute for Lifecycle
    Environmental Assessment
Ron L. Sass, Rice University
William J. Parton, Colorado State University
Steven M. Ogle, Colorado State University
Keith Paustian, Colorado State University
James Holtkamp, Holland & Hart LLP
Wiley Barbour, Environmental Resources Trust

Advisory and Review Committee

William H. Schlesinger (chair), Duke University
Michael Oppenheimer, Princeton University
Charles W. Rice, Kansas State University
Christopher B. Field, Stanford University
Steven Hamburg, Brown University

Harnessing Farms and Forests in the Low-Carbon Economy

How to Create, Measure, and Verify Greenhouse Gas Offsets

The Nicholas Institute for
Environmental Policy Solutions

Edited by Zach Willey
& Bill Chameides,
Environmental Defense

Duke University Press
Durham & London | 2007

© 2007 Duke University Press
Printed in the United States of America on acid-free paper ∞
Designed by Chris Crochetière, BW&A Books, Inc.
Typeset in Minion and Scala Sans by BW&A Books, Inc.

Library of Congress Cataloging-in-Publication Data
Harnessing farms and forests in the low-carbon economy:
how to create, measure, and verify greenhouse gas
offsets / edited by Zach Willey and Bill Chameides;
contributing authors, Gordon R. Smith . . . [et al.].
p. cm.
"The Nicholas Institute for Environmental Policy Solutions."
Includes bibliographical references and index.
ISBN-13: 978-0-8223-4168-0 (pbk. : alk. paper)
1. Alternative agriculture—Environmental aspects—
Evaluation. 2. Carbon sequestration—Measurement.
3. Carbon dioxide sinks—Measurement. 4. Emissions
trading—Evaluation. 5. Carbon cycle (Biogeochemistry)
6. Greenhouse gas mitigation. 7. Carbon dioxide mitigation.
8. Forest ecology. 9. Agriculture—Economic aspects.
I. Willey, Zach. II. Chameides, W. L. III. Nicholas Institute
for Environmental Policy Solutions (Raleigh, N.C.)
S589.75.H36 2007
363.738'746—dc22   2007014064

Contents

Appendices

Foreword

We have reached a critical point. Climate change is accepted by the scientific community as a real and present threat to our livelihood, one at least partly attributable to an increase in atmospheric greenhouse gases caused by human activity. To reverse this threat will require many arrows in the quiver. Foremost, a large-scale reduction of greenhouse gas emissions will be required across its many sources in our diverse economy. But these emission reduction efforts can also be bolstered by increasing the amount of carbon dioxide removed from the atmosphere and sequestered in terrestrial ecosystems. Therein lies the opportunity for the owners and managers of farmlands and forests to participate in the solution. Farmers can remove carbon dioxide from the atmosphere and sequester it as soil carbon by changing tillage practices; they can also modify agricultural practices to reduce greenhouse gases such as methane emitted in livestock and rice production, and nitrous oxide in soil management. Afforestation, or the planting of trees on currently nonforested lands, can transfer large volumes of carbon dioxide from the atmosphere to carbon storage in biomass, soils, and harvested products. Taken together, these activities can have a substantially favorable impact on the greenhouse gas balance. And if farmers and forest landowners can be compensated for their actions to reduce emissions or sequester greenhouse gases, they can benefit economically from these efforts.

Bringing farmers and foresters to the table shows great promise for mitigating climate change but requires a system that accurately measures and accounts for their greenhouse gas reductions. Because agriculture and land use activities are numerous and widely dispersed across a varying landscape, a well-designed system is necessary to bring order to the underlying complexity. This book builds the framework of just such a system by combining the insights of top scientists and economists in agriculture, land use, and forestry with the practical background of developers experienced in greenhouse gas mitigation projects, to systematize an approach that is scientifically grounded, has environmental integrity, and is practical to apply.

Bringing science into practice lies at the very core of the mission of the Nicholas Institute for Environmental Policy Solutions at Duke University. We believe that good science is the foundation of good environmental policy and practice. We also recognize that connecting science to practice often requires the type of hard work evident in the pages of this volume. For that we extend great thanks to all the technical experts who contributed to this effort and in particular to Zach Willey and Bill Chameides for coordinating, synthesizing, and communicating their work, clearly and logically. We believe that the collective effort constitutes a gold standard for including agriculture and forest activities in greenhouse gas offset programs.

This work would not have been possible without the financial support, enthusiasm, and vision of Peter Nicholas.

Timothy Profeta, *Director*
Brian Murray, *Director for Economic Analysis*
Nicole St. Clair, *Associate Director*
Nicholas Institute for Environmental Policy Solutions
February 2007

Preface

In 2003, a kind of economic nightmare seemed to be emerging in the United States. Although the nation had not created a mandatory cap-and-trade system for greenhouse gas (GHG) emissions, voluntary trading of such emissions and *offsets* (efforts to remove carbon dioxide from the atmosphere or prevent GHG emissions in the first place) had already begun. Businesses and individuals seeking to limit or neutralize their carbon footprint, their impact on global warming, began to purchase offsets from other businesses and individuals who had found ways to reduce their own emissions. Local markets and exchanges, brokerages, registries, and trading clubs sprouted up to meet the demand. However, the standards used to define the commodities to be traded varied wildly. In contrast, trade of GHG emissions and offsets among European Union nations was proceeding in a relatively orderly fashion. That is because participation in the Kyoto Protocol's cap-and-trade program had required the EU to create a regulatory framework with consistent and credible definitions of GHG offsets.

In the United States, a federal program regulating GHG emissions does not exist. The result is a piecemeal market for carbon offsets, in which the credibility of the commodities for sale can vary substantially. In the long run, this is an untenable situation for buyers and sellers alike. For buyers, *caveat emptor* ("let the buyer beware") is the watchword. For sellers, the lack of a system for verifying and validating offsets tends to depress the price they command.

Targeted changes in land uses and management practices in both agriculture and forestry can provide a major source of GHG offsets. These benefits result from using forests and soils to remove and store carbon already in the atmosphere and from reducing emissions of GHGs in the first place. The agriculture and forestry sectors have significant potential to help stabilize GHG emissions in the United States, particularly over the next several decades. For that to happen, however, such terrestrial GHG offsets must rest on transparent definitions and standards based on first-rate science. Such standards would give buyers and sellers alike a basis for establishing the value of the offsets and also provide a model for regulations that will surely ensue at the state and (eventually) federal level.

In early 2004, Environmental Defense contacted two groups of independent scientists to help provide these guidelines. The goal was to provide a gold standard for ensuring quality and integrity—a step-by-step guide to quantifying and verifying GHG offsets based on changes in land use and management in agriculture and forestry. Five highly regarded scientists agreed to serve on an advisory and review committee for the project. Dr. William H. Schlesinger, dean of the Nicholas School of the Environment and Earth Sciences at Duke University, chaired the committee. Dr. Schlesinger and his colleagues provided the wisdom and advice needed to steer this daunting, multidisciplinary project through its many technical mazes.

A second group of scientists then applied its unique

and varied experience to key aspects of creating terrestrial GHG offsets. These scientists contributed papers that answered the central question: how much will any specific farm or forestry project reduce levels of GHGs? Dr. Gordon R. Smith spearheaded the distillation of those papers into this guide, supported by the advisory and review committee and other consulting scientists. Dr. Dennis O'Shea, and later Sandra Hackman and Dr. Bill Chameides, then undertook two difficult tiers of editing.

All these individuals working in tandem over the past several years have produced the document that follows. We all are grateful to Peter Nicholas for his gracious funding—and infinite patience—in support of this work.

The extensive knowledge and guidance embodied here will provide invaluable direction to farmers, foresters, and other land managers, as well as consultants, brokers, investors, and others interested in creating consistent, credible GHG offsets as a new tradable commodity in the United States. This guide will help make tangible a new economic opportunity for rural America. In addition, it will provide important guidance to the policy community pursuing controls on GHG emissions—in the United States and other parts of the world.

Part I   **Overview**

Chapter 1

The Role of Landowners and Farmers in the New Low-Carbon Economy

A new economy is coming—a low-carbon economy in which greenhouse gas emission allowances and offsets will be a commodity that is bought and sold on the open market. Landowners and farmers, the people who work the land, will have a competitive advantage in this new economy because land, if properly managed, can be made to store carbon. Industries that emit carbon dioxide will pay landowners and farmers who store carbon to offset industrial emissions.

Why a Low-Carbon Economy?

The low-carbon economy will place a premium on technologies that can produce energy with little or no carbon dioxide (CO_2) emissions, as well as on activities that help remove carbon dioxide from the atmosphere. Why? The answer is simple: global warming. While uncertainties about climate remain, the basic facts of global warming are now well established:

- The globe is warming. The warming is due in large part to emissions into the atmosphere of CO_2 and other heat-trapping or greenhouse gases (GHGs) that result from human activities.[1]
- Unless we slow the rate of these emissions, the consequences could be dangerous, expensive, and irreversible.

In a communiqué issued in June 2005, 11 national academies of science (including the U.S. National Academy of Sciences) held that "the scientific understanding of climate change is now sufficiently clear to justify nations taking prompt action . . . We urge all nations . . . to take prompt action to reduce the causes of climate change."

The only way to curb human-induced climate change is to reduce emissions of CO_2 and other GHGs. And the only way to accomplish that is to move to a low-carbon economy that values technologies that limit GHG emissions and devalues technologies that produce GHG emissions.

Momentum toward a low-carbon economy is building. Thirty-five of the world's developed countries have agreed to reduce their GHG emissions 5 to 8 percent below 1990 levels through the Kyoto Protocol.[2] While the U.S. government has not joined the Kyoto process, many states and local governments have made Kyoto-like commitments. California has committed to a cap on its state-wide greenhouse gas emissions that will lead to substantial cuts in emissions in the coming decades. Four other southwestern states (Arizona, New Mexico, Oregon, and Washington) have joined California in the Western Regional Climate Initiative with the goal of setting a regional greenhouse gas emissions reduction goal. Nine northeastern states (Connecticut, Delaware, Maine, Massachusetts, New Hampshire, New Jersey, New York, Rhode Island, and Vermont) have joined the Regional Greenhouse Gas Initiative (RGGI) and agreed to cap CO_2 emissions from power plants. Many other states have announced climate initiatives and are considering statewide caps on GHG emissions.

In the private sector, major U.S. businesses (including Alcoa, BP America, DuPont, Caterpillar, and General Electric) have formed the United States Climate Action Partnership calling for mandatory caps on the nation's greenhouse gas emissions.[3]

Although the United States has yet to adopt a mandatory program to reduce GHG emissions, many people believe it is only a matter of time before it does. Indicative of this is a resolution passed in 2005: "It is the sense of the Senate that Congress should enact a comprehensive and effective national program of mandatory, market-based limits and incentives on emissions of greenhouse gases (S.AMDT.866)."

The Transition to a Low-Carbon Economy

History has shown that markets, rather than mandatory controls, can be the most cost-effective way to cut pollutant emissions. In a regulatory system, a market approach often takes the form of a "cap-and-trade" mechanism.[4] Such a mechanism caps total emissions from regulated entities—which may include a specific sector, such as power production in the case of RGGI, or the entire economy, as in the case of Kyoto—at a specified level, usually significantly below the current level. Regulators then assign individual emitters allowances, or caps, such that the total allowances equal the overall cap. Emitters have some period of time to comply with their cap.

Emitters can comply in three ways. First, they can use efficiency measures, technological advances, or lower activity levels to reduce their emissions. Second, they can purchase allowances from other emitters who have reduced their emissions below their caps. Third, they can purchase *carbon offsets* from individuals or entities, which remove CO_2 from the atmosphere or prevent GHG emissions.[5] This market approach allows emitters to find the cheapest way to meet their individual caps, as emitters that would incur relatively high costs can acquire allowances and offsets from those that can generate them at lower costs.

In this approach, CO_2 and other GHG emissions become a commodity that is bought and sold, and the marketplace (rather than regulators) determines the price of carbon allowances and offsets. These allowances and offsets can be relatively cheap or costly, de-

pending on supply and demand. Businesses and individuals also have an incentive to develop cost-effective methods of reducing GHG emissions and creating carbon offsets. By allowing the marketplace to control the price, the system guarantees that emitters will choose the most inexpensive and effective methods for reducing or offsetting emissions.

In unregulated systems, corporations and individuals can voluntarily cap their GHG emissions, as some companies have done. Cities and other municipalities have also adopted voluntary caps on the emissions arising from government activities. Voluntary caps usually do not include trading, but emitters may still purchase offsets when internal efforts to boost efficiency and adopt new technology do not produce the desired results. Here again the marketplace sets the price of the carbon offsets. As more companies and individuals take on a cap, demand for offsets rises, as does the price they command.

Despite the absence of a mandatory nationwide cap on GHG emissions, a U.S. market for carbon offsets is already burgeoning. Numerous companies have formed to buy and sell offsets, while other companies have emerged to verify and register those offsets. Many of these companies can be identified through a simple Internet search. However, potential buyers should exercise caution because the system is not yet regulated, and many developers of offsets do not yet follow rigorous procedures for creating them, such as those outlined in this volume.[6]

Farmers' Entrée into the Low-Carbon Economy: Carbon Offsets

Land-management practices can play a significant role in slowing the buildup of GHGs. Forests and farmlands act as natural carbon storehouses, or *sinks*, offering major opportunities to reduce global warming. As forests grow, they absorb CO_2 from the atmosphere, storing (or sequestering) vast amounts of carbon in wood, leaves, roots, and soils. Agricultural practices such as no-till or low-till farming, grassland restoration, and the use of cover crops also sequester carbon in soils. By protecting and restoring forests, replanting grasslands, and improving cropland-management practices, landowners can help reduce atmospheric concentrations of GHGs.

Besides removing carbon already released into the atmosphere, better land-use practices can also reduce emissions of potent GHG such as methane and nitrous oxide. For example, using fertilizer more precisely can reduce emissions of nitrous oxide from soil. Reducing the saturation of soil with water (particularly during rice cropping) can curb methane emissions, as can the capture and burning of methane emitted from manure.

While environmentalists have pointed to the potential for these activities to slow global warming, farmers and landowners today have little economic incentive to adopt them. However, this will change as the transition to a low-carbon economy puts a market value on land-management practices that store carbon and reduce GHG emissions.

In fact, even where caps on emissions remain mostly voluntary, offset projects targeting carbon dioxide, methane, and nitrous oxide are already under way. In the Northwest, the energy company Entergy has funded Pacific Northwest Direct Seed Association, a nonprofit composed of more than 100 farmers, to create marketable offsets by using low-till farming to sequester carbon in soil and lower CO_2 emissions. In the Midwest, a grain-milling cooperative is creating offsets based on the land-management practices of several hundred farmers in Kansas, Missouri, Nebraska, and Iowa, such as the use of no-till farming to store more carbon in soil. In the Northeast, a group of dairy farmers is seeking buyers for offsets based on cuts in methane emissions resulting from the use of anaerobic digesters to treat manure. In the South, a consortium of farming operations is creating offsets by shifting to low-till cropping to reduce CO_2 emissions, changing crop rotations to store more carbon, and improving livestock and manure management to reduce methane emissions.

The Potential of Offsets Based on Land Management

Land-management practices have the potential to make a significant dent in GHG emissions. The U.S. Envi-ronmental Protection Agency (EPA) estimates that the United States emits some 6,000 million metric tons of CO_2 each year, as well as the equivalent of another 1,000 million metric tons of CO_2 in the form of other greenhouse gases, including methane, nitrous oxide, and chlorofluorocarbons. Overall, annual GHG emissions total the equivalent of some 7,000 million metric tons of CO_2 (see Figure 1.1).

If the United States takes no steps to reduce GHG emissions, how large would they be in, say, 2025? The recent past can provide a clue. In 1990, U.S. greenhouse gas emissions were equivalent to about 6,100 million metric tons of CO_2 per year; in 2004, they were reaching nearly 7,100 million metric tons. GHG emissions are therefore rising at an annual rate of about 1 percent. Without a limit on such emissions, we can assume they will continue to rise an additional 1,600 million metric tons per year by 2025, to the equivalent of about 8,700 million metric tons of CO_2 annually.

Climate models suggest that by the later part of the twenty-first century, humanity must reduce global GHG emissions by about 50 percent from their present rates to avoid dangerous climate change (O'Neill and Oppenheimer 2002; Den Elzen and Meinshausen 2005).[7] This prospect is challenging to say the least. In the United States, this would require cutting annual emissions by some 3,500 million metric tons of CO_2. The good news is that we do not have to attain this 50 percent reduction immediately. We can slowly ramp down our emissions to reach the 50 percent reduction by the end of the century, when new technologies and energy sources will hopefully have replaced the carbon-intensive forms we rely on today.

Over the next 20 years or so, developed nations might reasonably aim to lower their emissions by about 15 percent (Den Elzen and Meinshausen 2006). For the United States, this would require cutting the equivalent of about 1,000 million metric tons of CO_2 per year. Adding the estimated annual increase in GHG emissions during this period of 1,600 million metric tons, the United States would have to find emissions cuts equivalent to about 2,600 million metric tons of CO_2 per year. Although not as imposing as the 50 percent target, this goal will still significantly test our economic and technological ingenuity.

Could land-management practices help the United States meet the 20-year target cut of 2,600 million metric

Figure 1.1 U.S. CO_2 and other greenhouse gas emissions, 1990–2004 (in millions of metric tons of CO_2 equivalent). Emissions rose at an average annual rate of about 1% over the period. If that rate persists, U.S. emissions will grow from the present 7,000 million tons a year to about 8,700 million tons in 2025. *Note*: From U.S. EPA 2006.

Figure 1.2 Carbon offsets that U.S. land-management practices could create, as a function of year and price (in millions of metric tons of CO_2 equivalent). With the rising price of offsets the total amount of offsets available should increase, as more farmers and landowners perceive an opportunity to profit and participate in the market.
Note: From U.S. EPA 2005.

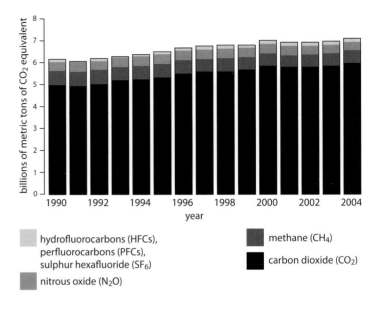

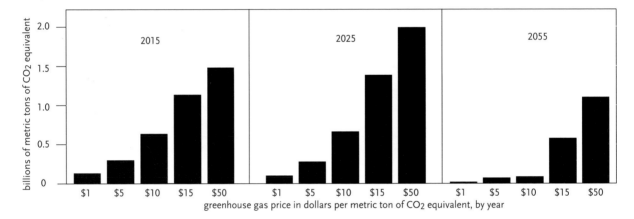

tons of CO_2 per year? Consider a recent EPA study (2005), which estimated the potential for carbon offsets from land-management practices (see Figure 1.2).[8] Not surprisingly, as the price of offsets rises, more farmers and landowners opt to participate in the market, and thus the total amount of offsets also increases. The amount of offsets also depends on time. Although the amount of offsets grows as more farmers and landowners participate and soils and forests increase their capacity to store carbon, the amount of offsets could peak in 2025 because soils and forests eventually become saturated with carbon and lose their ability to store more. The amount of offsets could even decline if cutting of forests used to create offsets outstrips reforestation.

The results from the 2005 EPA study suggest that land-management practices can play a major role in enabling the United States to meet the emissions target over the coming decades if the price of carbon offsets is high enough. If offsets command a price of \$15 per ton of CO_2, land-management projects could offset almost 1,500 million metric tons of CO_2 per year by 2025—around 60 percent of the needed reduction. At \$50 per ton, offsets could total almost 2,000 million metric tons of CO_2 per year—nearly the total required cut in emissions.

Will the price of offsets be high enough to generate the needed amount? That depends on demand. In the United States, where emissions caps are voluntary and the market for offsets is currently relatively weak, off-sets are now selling for a few dollars to about $10 per ton of CO_2. However, in the European Union, which has adopted a mandatory cap under the Kyoto Protocol, CO_2 prices rose into the range of $30 to $40 per ton of CO_2 in 2006. This suggests that if the United States adopts a mandatory cap, the price for offsets will be high enough for land-management practices to play a major role in meeting the cap. Because carbon offsets will be critical in the transition to low-carbon technologies, farmers and landowners who enter the offset market early stand to profit the most.

The Need for Offset Quantification Guidelines

While projects based on changes in land-management practices have the potential to offset significant amounts of GHG emissions and to provide a new income stream for farmers and landowners, they present significant challenges to the individuals and entities that undertake them. At the front end of an offset project, developers need to reliably estimate its potential value and thus the amount of GHG mitigation it is likely to produce. As any farmer can attest, projecting crop yields at the beginning of a planting season is difficult. In an offset project based on changes in land management, developers must attempt to project outcomes over many years, in some cases more than a decade. Moreover, to market the GHG mitigation they achieve, project developers must reliably document it. This, in turn, requires developing and implementing a comprehensive plan for monitoring and analyzing the results of the project, as well as contracting for independent verification of the plan and its implementation.

Monitoring itself presents challenges. Instead of simply documenting the yield of wheat or corn, land managers must quantify the amount of carbon they store in soil or forest wood or the amount of methane they capture from processed manure. To ensure that the project does in fact lead to real GHG benefits, land managers must also often track conditions and carbon-sequestration rates on nonproject lands. They must make a long-term commitment to monitoring and tracking. Not only does the amount of carbon a project adds to soil or forest vary from year to year, but

the carbon stored in years past can be lost because of fire or annual changes in climatic conditions. Finally, marketing carbon offsets requires careful analysis of monitoring and tracking data to ensure that the offsets claimed are accurate with a known and acceptable level of uncertainty.

An additional complication arises from the fact that the validity of any carbon offset project is ultimately based on our scientific and technical understanding of how carbon and other elements are cycled through agricultural and forest systems and how these systems interact with the climate system. Because science is continuously evolving, the system used to manage, quantify, and verify the value of a carbon offset project must be sufficiently flexible to accommodate scientific advances. See for example Keppler et al. (2006), Gibbard et al. (2005), and Olander (2006).

Furthermore, for buyers, regulators, and the public to accept offsets stemming from changes in land management, they must have confidence that the mitigation is real. Credible and transparent rules and methods are therefore critical to ensure that offsets are fully tradable. This volume attempts to address this need by providing specific guidelines for developing and implementing land-management projects that produce carbon offsets.

This Manual

This manual aims to provide a comprehensive, user-friendly description of the principles and methods needed to quantify cuts in GHG emissions and removal of CO_2 from the atmosphere stemming from land-management practices. These principles and methods build on years of scientific study of the most accurate ways to measure changes in methane and nitrous oxide emissions from soil and manure and changes in carbon stocks in trees and soil. The approaches presented here aim to strike a balance between reliability and affordability. That is, participants in the system, regulators, and the public must believe that the offsets landowners create are real, but the costs of measuring and verifying the offsets must not rise so high that projects become economically impractical.

Types of Projects

This volume focuses on four basic categories of land-management projects designed to create marketable carbon offsets:

1. Projects designed to sequester carbon in soils, such as through the adoption of no-till farming.
2. Projects designed to sequester carbon in biomass through cultivation of new forests and grasslands or delays in harvesting forests.
3. Projects designed to reduce methane emissions through changes in the practices used to process and dispose of manure.
4. Projects designed to reduce emissions of methane and nitrous oxide through changes in farming practices.

Farmers and landowners also have other options for developing carbon offsets, such as by producing bio-energy crops and constructing wind turbines for generating power. However, because these types of projects do not involve specific land-management practices, this volume does not address them.

The Audience

This book is designed for use by all who might participate in developing, marketing, and purchasing offsets based on changes in land management. These include:

- *Landowners*, on whose land a project is executed.
- *Farmers*, who pursue project activities.
- *Project developers*, who plan and implement the project, even though they may or may not be the farmers or owners of the land.
- *Quantifiers*, who perform the monitoring and analysis required to assess the quantity of legitimate offsets the project achieves and who may or may not be the project developers.
- *Verifiers*, independent agents who audit the quantification of the project's offsets, vouching for their accuracy and adherence to specific guidelines established by regulators of a carbon market.
- *Regulators*, who develop and enforce regulations governing carbon offsets in a cap-and-trade system.

- *Retailers or brokers*, who may purchase offsets from multiple projects, aggregate them, and resell them directly to buyers or through a carbon offset market.
- *Buyers*, who purchase offsets directly from project developers or retailers or through a carbon offset market.
- *Offset owners*, who have legal ownership of offsets and who may be the landowner, project developer, retailer, or ultimately the buyer.

Landowners, project developers, quantifiers, regulators, and retailers are obviously interested in the principles and methods needed to produce accurate and credible offsets. However, buyers of offsets would also be well advised to understand the basic principles used to produce offsets because creating them can be challenging, and potential buyers, especially in unregulated markets, need to assure themselves that the offsets they purchase are real. For example, some carbon offsets for sale in the United States have not been independently verified, and others lack evidence that they represent GHG benefits that would not have occurred without the project. Those projects that adopt the principles and methods outlined here should not be subject to these types of shortcomings.

Applications of the Manual

This volume could be valuable in at least three scenarios involving the development of carbon offsets:

1. Voluntary development on the part of landowners without a carbon offset market: This scenario does not involve a mandatory, government-imposed cap-and-trade program. Instead, landowners who want to voluntarily offset their emissions embark on a project.
2. Voluntary development by individuals and companies within a carbon market: Although regulators have not imposed a mandatory cap-and-trade program, individuals and companies who want to voluntarily offset their emissions contract with landowners and developers or retailers to purchase offsets. This situation now applies to most of the United States.

3. Mandatory development for major emitters within a government-imposed cap-and-trade program and carbon market: This situation now applies to power companies participating in the Northeast's (U.S.) Regional Greenhouse Gas Initiative and to countries participating in the Kyoto Protocol.

This manual is primarily targeted to the second and third scenarios. Of course, any regulatory systems that limit GHG emissions and allow trading will require the use of specific procedures to create offsets. Such systems may also accept only certain types of offsets greater than a specified size, and they likely would require authorized entities to quantify them.[9] In these cases, the regulatory system's guidelines will supersede those presented here. However, even in such cases, this manual should prove useful in helping individuals interpret and understand regulatory requirements. This volume also can serve as a guide to legislators and regulators who aim to design, implement, and strengthen a cap-and-trade system that includes land-management options for offsetting GHG emissions.

The Organization of the Manual

This manual provides a comprehensive overview of the principles that underpin carbon offsets based on changes in land management, as well as the methods used to quantify them. It is divided into three sections. The first provides an overview for legislators, landowners, and those who are unfamiliar with offset markets but interested in learning about them. The second provides a more detailed but nontechnical exposition of the offset process for project developers, investors, and purchasers of offsets. The third, contained in the appendices at the end of the volume, provides the technical information that is critical to the individuals responsible for quantifying, verifying, and/or regulating offset projects.

Chapter 2

The Process of Creating Offsets

To qualify as a marketable carbon offset, the greenhouse impact of a change in land management must have three critical attributes:

1. It must represent a net reduction in GHG emissions or a net gain in the amount of carbon stored in soil, trees, or other biomass, compared with what would have occurred on the land without the project.
2. The offsets must have a legal and specified owner. Depending on the contractual arrangement and whether the offset has been marketed, the owner may be the landowner, the project developer, a retailer, or a buyer.
3. Regulators of any relevant cap-and-trade system and the buyer, as well as the public, must have strong confidence that the offsets have been accurately measured and quantified.

By convention, offsets are expressed in tons of CO_2 equivalent (CO_2e), reflecting the global warming potential of different greenhouse gases.[1]

Greenhouse gases are invisible to the naked eye, and buyers of offsets do not physically take delivery of tons of gas the way buyers take possession of bushels of corn. Thus, before committing to a project, landowners and buyers alike will want reasonable assurance that it will provide the offsets they seek, with understood and acceptable levels of uncertainty and risks. To obtain such assurance and to guarantee that the offsets

a project produces are real, participants must navigate a complex series of steps (see Figure 2.1). This chapter provides an overview of those steps; more details are presented in subsequent chapters.

Defining a Project

Landowners and project developers should begin by defining the land-management practices they will use to create offsets. For example, a project might entail sequestering carbon in soil or trees or using anaerobic digesters to capture methane emissions from manure. Project developers must also establish the project's spatial and temporal boundaries so they can quantify and verify any carbon offsets the project creates. Project developers can use land surveys to specify a project's boundaries. However, this can often prove costly, and developers may opt for less-expensive options, such as relying on planning maps or a GPS receiver to record project boundaries. Developers also can rely on legal records of land parcels or a suitably labeled and marked aerial photograph. Some landowners may be tempted to use roads or rivers to delineate project boundaries. However, seemingly permanent landmarks shift with surprising frequency, as roads are relocated and floods move river channels, so this approach is impractical. Project boundaries may be highly irregular or discontinuous, such as when a project entails planting trees on a sinuous floodplain. In that case, project developers may use the anticipated extent of activities to plan

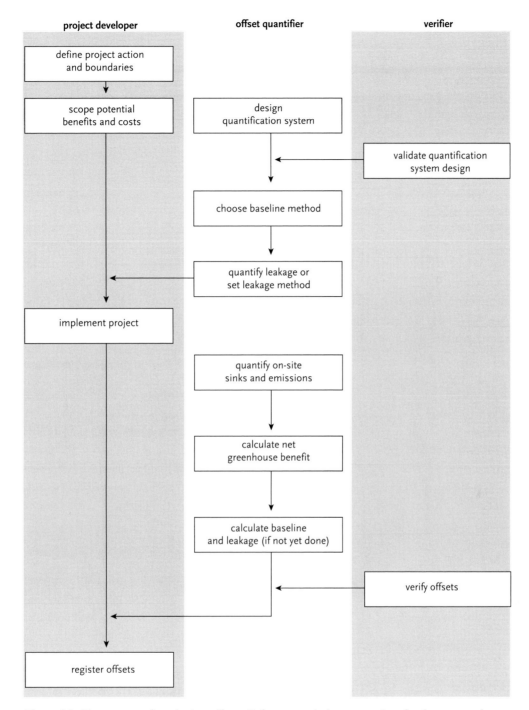

project developer　　　　**offset quantifier**　　　　**verifier**

define project action
and boundaries

scope potential
benefits and costs

design
quantification system

validate quantification
system design

choose baseline method

quantify leakage or
set leakage method

implement project

quantify on-site
sinks and emissions

calculate net
greenhouse benefit

calculate baseline
and leakage (if not yet done)

verify offsets

register offsets

Figure 2.1 The process of producing offsets. Before committing to a project, landowners and buyers alike will want reasonable assurance that it will provide the offsets they seek. To obtain such assurance, participants must navigate a complex series of steps.

the project, but use monitoring data gathered during the project to determine the areas where project activities actually occur. For projects located at a specific facility, such as a farm with animal barns and associated manure-handling areas, an address may be adequate to specify the project's spatial boundaries.

If landowners have diverse operations, project boundaries should encompass all the lands and facilities related to the project. Otherwise an owner could create offsets simply by shifting activities that produce GHG emissions to lands or facilities outside the project. For example, consider a meat company that grows crops, uses them for feed, slaughters animals, and manufactures food products. If the company aims to reduce its methane emissions (and thus create carbon offsets) by capturing methane from decaying manure, it would clearly need to include the facilities used to process the manure within the project boundary, and it could reasonably exclude the slaughterhouse and manufacturing operations. However, the boundary should include feedlots as well as lands where the manure is spread as fertilizer after processing. That is because the decaying manure in these locations may produce unwanted methane emissions, which quantifiers of the offsets must take into account.

Offset projects are bounded in time as well as space. Time boundaries are usually relatively easy to specify. Projects typically start on the date a contract takes effect, a land-management activity begins, or a facility or process (such as capture of methane from animal waste) starts operating. Projects may run indefinitely, or a contract among landowners, any outside investors, and buyers may specify an ending time or process.

Scoping a Project's Costs and Benefits

After establishing the boundaries of a project, developers should scope it; that is, they should estimate the greenhouse benefits they expect it to produce, to see if it will create offsets, and estimate its costs, to see if it will yield a net financial return. This process enables developers to decide whether to implement a project. A project's costs include those of implementing it; creating contracts among participants; and measuring, verifying, registering, and marketing the resulting off-

sets. In assessing the net financial return, landowners, developers, and buyers need to consider at what point during the project's lifetime the costs and GHG benefits will occur.

Estimating a project's potential for producing offsets entails a number of steps that are conceptually the same as or similar to those used to actually quantify the offsets. However, at this stage, developers use less rigorous and therefore less time-consuming and costly methods, such as estimates and projections, instead of actually monitoring project outcomes. Project developers pursue each of these steps in the scoping phase, whereas quantifiers perform them during the actual project. The following sections outline the principles that underpin each step; later chapters provide more detail on specific methods for quantifying the GHG impact of offset projects.

Establishing Additionality and Baselines

In order for the offsets claimed by a project to be valid, the changes in GHG emissions or carbon stocks on project lands must be different than those that would have occurred in the absence of the project, and the difference must be quantifiable. Project developers can address those issues by determining *additionality* and *baselines*.

Deciding Whether a Project Is Additional

Valid and marketable carbon offsets must be *additional*. That is, any reductions in GHG emissions and increases in stores of carbon produced by the project would not have occurred without the project.[2] Determining whether a project is additional entails finding sites with similar starting conditions and anticipating changes in land management likely to occur on those lands during the lifetime of the project. This will determine whether the project will indeed produce GHG savings over and above "business as usual."

Although determining whether a project is additional is challenging, that effort is critical. If a project's GHG benefits are not additional, then the offsets the project claims are not real, and a cap-and-trade system should not credit them. For example, consider an electric power plant whose owners would like to exceed its cap of 1 million tons of CO_2 per year, emitting 1.1 million tons. To comply with the cap, the company buys 100,000 tons of GHG offsets from a farmer who has sequestered extra carbon in the soil by switching from plowing to no-till cultivation. If the farmer planned to take that step for reasons other than sequestering carbon, the utility would still in effect be emitting 1.1 million tons.

Setting a Project's Baseline

The GHG emissions from a project's lands and facilities that would have occurred in the absence of the project constitute its *baseline*. The baseline often changes over time due to changing management or environmental conditions, including climate change. The project's net GHG benefit is the difference between the baseline and the actual GHG emissions from lands and facilities during the project.

If a facility such as a methane digester captures GHG emissions, measurements of emissions in the absence of the digester—that is, before the project began—can serve as the baseline. Other types of projects must establish the baseline as they unfold, to account for changing conditions and land-management practices on lands with similar conditions in the region.

Some systems, including the Kyoto Protocol, treat additionality and baselines as separate steps. The system establishes whether a project is additional, and if it is, the system specifies an independent method for determining the baseline. This process usually sets an all-or-nothing test for additionality. The Kyoto Protocol provides an example of this type of test. That regime views a land-management practice as additional if certain tests are met (e.g., if less than some specified percentage of farmers in the region have already adopted the practice). If these tests are met, a project's net GHG benefits are deemed 100 percent additional and therefore allowed as offsets. If not, the project is deemed categorically nonadditional, and the system will not accept any of the project's net GHG benefits as offsets.

However, this all-or-nothing approach tends to discourage the increased use of practices that actually reduce GHG emissions, and thus it can be counterproductive. For example, assume that a farmer seeks to sequester carbon (and thus produce offsets) by switching from plowing to no-till farming of corn and soybeans. Farmers in the region use no-till methods on 60 percent of the land planted in a corn-and-soybean rotation. Systems using a strict all-or-nothing form of additionality, such as the Kyoto system, would disqualify new projects based on no-till farming, even though those projects would sequester more carbon.

We propose a different methodology, often referred to as *proportional additionality*, which considers additionality and the baseline simultaneously. The project developer must first identify comparison lands that are similar to the project lands. The project's baseline is then assumed to be the emissions or changes in carbon stocks on the comparison lands during the project. Because land-management practices largely determine the emissions or changes in carbon stock on comparison lands, this method implicitly accounts for additionality, but in a proportional way rather than an all-or-nothing manner. For example, suppose the land-management practices on comparison lands are the same as those on project lands; the project is not additional at all. In this case, all other things being equal, the baseline would be the same as the emissions or changes in carbon stocks on project lands, and thus the project would not produce any offsets. If, on the other hand, half of the comparison lands used the same management practices as the project lands and half did not, developers would reduce the amount of net GHG benefits they produce by half, in keeping with the fact that the project is 50 percent additional.

Quantifiers must express the baseline in the same units they use to quantify changes in GHG emissions and carbon stocks on project lands, typically, in tons of CO_2e. If quantifiers calculate the baseline on a per-acre (or per-unit) basis, they must multiply the result by the number of acres (or units) encompassed by the project. If the project reduces GHG emissions, then quantifiers subtract baseline emissions to obtain the project's net GHG benefit. If the project sequesters carbon, quantifiers subtract baseline carbon stocks to find the project's net benefit.

Project developers typically divide a project into accounting periods during which they quantify and market offsets. For example, a project that lasts 10 years might quantify its net GHG benefits for 10 one-year accounting periods. Quantifiers must determine the baseline for each accounting period.

Creating a Monitoring and Quantification Plan

To ensure that a project's offsets are verifiable and thus acceptable to regulators and buyers, developers must use documented and transparent methods to create and quantify them. They must therefore write a monitoring and quantification plan before beginning a project. If changing conditions require revisions in the project's land-use or quantification practices, addendums to the plan should document these revisions. Before developers can market offsets generated from the project, these offsets must be independently verified. Verifiers will refer to the documented plan to ensure that developers have implemented the planned activities and that quantifiers have calculated net GHG benefits reliably.

Each project's monitoring plan should be concise but complete enough that others can comprehend it and accept offsets based on it. (A public version of the plan may contain less detailed information about management of project lands and their production.) The monitoring and verification plan should include the following:

- Project boundaries if they are known when the plan is written.
- A list of the project's potential sources and sinks of GHG emissions.
- Detailed protocols for measuring project conditions, including the frequency of measurements and procedures for recording, managing, and storing data. The plan should include a design for sampling the project's GHG impacts, including the locations of monitoring activities and mechanisms for identifying those locations, such as markings on aerial photos, GPS coordinates, or physical markers.
- Procedures, factors, and equations for analyzing data (based on measurements from the project) and quantifying offsets.

- Contents and timing of reports that quantifiers will generate. These reports should include summaries of original data, not just the results of data analysis, so someone unfamiliar with the project can interpret the information and analysts later can apply new techniques as they become available.
- Quality-control standards and methods, including redundancy in recording data in case records are lost.
- A baseline or a process for setting the baseline.
- Leakage rates or a process for assessing leakage (see below).

Designing Measurement Plans

The most important part of the monitoring and quantification plan specifies the methods used to monitor changes in GHG emissions or carbon stocks that result from the project's land-management practices. Developing such a plan entails many assumptions and professional judgments. To establish that emission cuts or greenhouse sinks are real, quantifiers must use techniques that are reliable, appropriate, complete, unbiased, and transparent. Critical in this regard is a careful sampling design for gathering representative and accurate data that can detect actual changes in emissions or sinks. (See Appendix 1 for key aspects of such a sampling design.)

Different types of land-management projects require different techniques for measuring net changes in emissions and sinks during the project. Individual projects address various components of ecosystems, and quantifiers must measure each component using techniques appropriate to it. Designing complete and accurate measurement systems requires a thorough understanding of the biological and physical attributes of what is being quantified. If not based on knowledge of such dynamics, a measurement system may not address all the GHG flows that a project is likely to affect. (See Chapters 6, 7, 8, and 9 for specific techniques and methods appropriate for each type of project and flow.)

The accuracy of claims for a project's offsets will generally reflect the rigor of the methods used to measure and quantify them. This, in turn, means that project developers face a quandary. A more rigorous measurement plan will produce more accurately quantified offsets, but it will also drive up the cost of producing them. On the other hand, a less-rigorous measurement plan will lower project costs, but it will also drive down the price the offsets can command in the marketplace. Successful project developers will strike a balance between rigor and cost-effectiveness, producing real GHG benefits while also realizing an acceptable profit.

For example, in theory, it is preferable to measure all GHG flows and carbon stocks of the project as directly as possible. However, this can become prohibitively expensive. Less-direct methods, such as models and empirical relationships, tend to yield less accurate results. One reasonable approach is to carefully measure major GHG flows and carbon stocks, while relying on models and empirical relationships to estimate lesser flows and stocks that will have a minor impact on the amount of offsets a project creates.

Deciding Which Emissions and Sinks to Quantify

Along with establishing a project's baseline, quantifiers must assess all potentially significant changes in GHG flows within the project's boundaries. These flows include inadvertent emissions arising from activities related to the project, such as from the burning of fossil fuels to operate farm equipment. (See Appendix 2 for more on inadvertent emissions and how to quantify them.)

However, quantifiers do not normally have to account for "upstream" and "downstream" emissions related to the project because project developers do not control those emissions. Upstream emissions result from the production of inputs to a project. For example, tree-planting projects may require seedlings grown in a nursery, which may use vehicles that burn fuel that emits greenhouse gases. The nursery can choose higher- or lower-emission methods of growing seedlings. However, the tree-growing project does not control those choices, and thus quantifiers should not have to include them in the project's GHG inventory. On the other hand, quantifiers should take into account emissions associated with transporting and planting the seedlings.

Counting upstream emission cuts can lead to dupli-

cation. For example, if a new forest plantation grows wood fiber to replace concrete or steel in construction, the landowner, the manufacturer of the wood products, the contractor who builds buildings, the owner of the buildings, and the steel or cement manufacturer may all want to claim cuts in greenhouse gases. However, one ton of emission cuts can legitimately offset only one ton of emissions. These types of conflicts are resolved by allowing those directly responsible for emissions cuts and carbon offsets to own them. In this example, the landowner could not claim the emissions cuts. Instead, regulators of a mandatory cap-and-trade system would likely allow the steel or cement manufacturer to claim an emissions credit.

Downstream emissions occur when other parties act on a project's outputs. For example, a farmer may sequester carbon in soil by reducing tillage. That farmer might produce corn, and the buyer of the corn might use it to feed cattle, which produce methane when digesting the grain. The farmer who grows and sells the grain does not have to take responsibility for the methane emissions of someone else's cattle because the farmer does not control those cattle.

However, a project should account for downstream GHG emissions from lands and waters owned by others if those emissions result directly from project activities and not from anyone else's activities. For example, some nitrogen fertilizer applied to fields often ends up in streams, where a portion turns into nitrous oxide that is emitted to the atmosphere. The leached nitrogen is a side effect of producing crops, not a marketed product of the farm operation. Because no other downstream landowner is causing the emissions, a project that uses fertilizer must take responsibility for those emissions. (Most projects would report average emissions from N_2O leaching in the region, rather than attempting to actually measure those emissions.)

Some landowners aiming to create offsets may be tempted to take credit for upstream cuts in emissions to which they are not entitled. For example, if a farmer uses less chemical-nitrogen fertilizer, he or she may wish to claim credit for a drop in emissions from the manufacture of the fertilizer. However, because the upstream manufacturer of the fertilizer is not participating in the project, the farmer does not own those cuts and thus would not be able to sell them as GHG offsets.

Creating Contracts for Trading Offsets

Because offsets are property, property law determines how landowners, project developers, and buyers can transfer ownership of offsets. For this reason, participants normally complete a contract before a project begins that clearly articulates the nature of the project and allocates costs, risks, liabilities, and profits.

Contracts between landowners and project developers may commit the former to performing specified activities or delivering a certain number of offsets. Contracts between project developers and buyers usually specify delivery of a certain number of tons of sequestered carbon or cuts in greenhouse emissions, calculated using a specified process. In either case, the contract must specify who owns the rights to the offsets and, if applicable, prohibit the landowner from registering or otherwise claiming GHG benefits, to reduce confusion and the risk of double-counting. To be legally valid, contracts must provide some benefit to each party, such as allocating carbon offsets to one party in exchange for payments to the counterparty.

Contracts should include a project schedule. For example, a contract governing sequestration of carbon in trees or soil should refer to a monitoring and verification plan, if completed, or a schedule for developing and approving such a plan. The contract should also include a schedule for implementing the plan over a specified period. Even if a contract covers only the transfer of existing offsets, it should specify dates for payments and delivery of offsets.

Contracts that transfer rights to offsets based on emissions cuts, such as those based on capture of methane from manure, may run for relatively short periods of time, such as one to five years. Contracts that transfer offsets based on sequestering carbon, in contrast, usually run for at least a decade.

Carbon-sequestration contracts may involve a perpetual commitment to keep carbon stored, or they may run for only a fixed period. If the commitment to store carbon is perpetual, it cannot be secured only by a contract. Contracts, by definition, can run for only a limited period of time. A buyer can ensure a perpetual obligation only by obtaining a property interest from the seller, usually in the form of an easement requiring or prohibiting certain land uses or management practices. (An easement should be recorded with the title to the land, so new landowners cannot claim they did not know about it.) However, many landowners may be unwilling to grant a perpetual easement for much less than the price they would gain from selling the property outright. In that situation, a project developer may wish to buy the land rather than an easement. As an alternative, the buyer or developer could obligate a landowner to maintain carbon stocks for a specific period of time (at a reduced cost), with the expectation to negotiate a new contract with the same landowner or a different landowner when the original contract expires.

Offset contracts vary widely in the way they distribute risks. For example, a project that requires a significant capital investment (such as one involving planting trees for forest sequestration or building a manure digester for reducing methane emissions) may specify payments from buyers long before the project creates offsets. In return for advance payments, a buyer may require a security interest in lands or facilities, to limit the risk that the project may not produce the offsets. One approach is again to record a lien, covenant, or easement on the title to the land. This interest can expire at a certain time or after the buyer acknowledges that the seller has fulfilled the contract's obligations.

Contracts may specify delivery of a fixed number of tons, or they may require (or prohibit) specified activities and transfer all resulting tons to the buyer. If the contract obligates the landowner only to perform specified activities, the buyer bears the risk that the activities will not generate the anticipated amount of offsets.

Besides specifying the amount of offsets that a project will provide, a contract may also warrant that an independent verification organization (in the case of a voluntary program) or a specified regulatory agency (in the case of a government-mandated cap on GHG emissions) will accept the offsets. To satisfy buyers who must comply with emissions caps, contracts may guarantee that a project developer will deliver offsets of comparable quality if the anticipated offsets do not appear or are deemed unacceptable. As an alternative, buyers may opt to purchase insurance from a company that promises to provide suitable offsets if regulators or independent verifiers do not accept the contracted offsets. Contracts may also specify the steps that the parties will take if one claims default and the other contests the claim. A common procedure is to require arbitration with specific rules for that process, including how to select and pay the arbitrator. Parties commonly split the cost of arbitration, especially if it will define a settlement rather than assign fault.

Parties wishing to transfer ownership of offsets may consult several sources for examples of contracts:

- The Master Agreement for the Purchase and Sale of Emission Products, by the Environmental Markets Association, covers offsets sold outside an emissions cap (http://www.environmentalmarkets.org).
- The International Emissions Trading Association (IETA) offers multiple model contracts for transferring offsets within the European Trading System to comply with the Kyoto Protocol (http://www.ieta.org).
- The Emissions Trading Master Agreement for the EU Scheme V2.0 provides a framework for trading offsets between two parties or for multiple trades of emissions allowances.
- The IETA's Clean Development Mechanism Emission Reduction Purchase Agreement V2.0 is specifically tailored to trading offsets before they are created. This detailed contract, which includes provisions for verifying offsets under the Kyoto Clean Development Mechanism process and for dealing with a project that produces fewer or more offsets than anticipated, can serve as a checklist for shorter contracts.
- The Climate Trust website (http://www.climatetrust.org) describes the terms the trust will ask for in a contract to purchase offsets.
- *New Tools for Improving Government Regulation: An Assessment of Emissions Trading and Other Market-Based Regulatory Tools* by Gary Bryner, published by the Center for the Business of Government, provides a basic primer on offset contracts.

Of course, parties must modify model contracts to fit particular transactions and consult legal counsel on the appropriateness of particular provisions, given the laws of the locality that governs a contract.

The farmer could take credit, on the other hand, for the drop in N_2O emissions from project lands that results from the use of less fertilizer.

To make accounting more tractable, policy makers may specify that emission mitigation be counted either upstream or downstream from where it is actually occurring. For example, a building constructed of wood keeps carbon stored in its wooden structure. However, it would be very expensive to quantify and track the wood in hundreds of millions of structures. Policy might assign the right to count the emission mitigation provided by holding carbon in wooden structures to landowners who grow timber. If the right to count the carbon storage in wood products is assigned to the grower of the tree these products are made from, then the counting is done upstream from the physical storage of the carbon in a structure. The key is that double counting is avoided because policy establishes a single point in the production chain at which a particular kind of GHG benefit is counted.

Greenhouse gases removed from the atmosphere and re-emitted within the same accounting period do not count as offsets. For example, if a farmer grows and harvests a crop but then feeds the grain to livestock and allows the residue to decompose, the project would produce no offsets. The CO_2 from animal respiration and the decay of plant residue replaces the CO_2 initially removed from the atmosphere by the growing crop.

Project developers do not create new offsets by simply maintaining those they previously created. For example, suppose that in the first year of a project, trees absorb and store one ton of CO_2 from the atmosphere, and the project gains credit for this as an offset. After the first year, the trees are stricken by drought and stop growing. The project will not create no new offsets after the first year, as it simply continues to store the carbon sequestered and credited in year one.

Determining Who Owns Offsets

If carbon offsets are traded in markets, they are legally a form of property, and their owners must be clearly and uniquely identified. Therefore participants in a project (whether landowners, project developers, or buyers) should assess ownership of any offsets. This normally requires assigning the rights and responsibilities for producing and owning them (see sidebar on "Creating Contracts for Trading Offsets").

The owner of lands or facilities that reduce GHG emissions or sequester carbon usually has a strong claim to the resulting offsets. However, in practice, multiple parties may have claims to carbon offsets. For example, a farmer renting a field may sequester carbon in the soil by switching from plowing to no-till cultivation. However, the landowner—not the farmer—owns the soil storing the carbon, and that landowner may later pursue land-use practices that release those stores, such as by reinstating plowing. A prudent buyer of carbon offsets would require both the farmer and the landowner to sign a contract assigning rights to any offsets and would require both to adhere to land-management practices that protect those offsets.

Unclear or contested land titles weaken rights to offsets unless all parties with claims to the land contractually assign the rights to any offsets. If landowners have previously enrolled their lands in the Conservation Reserve Program or another federal program that requires them to stop farming and plant grass or trees, the federal government may own the offsets that result because the government compensates farmers for participating in the program. However, in a cap-and-trade system, policymakers may choose to transfer ownership of any resulting offsets to landowners to provide another incentive for them to participate.

Buyers typically prefer to purchase large blocks of offsets instead of contracting with many landowners and project developers for numerous small and independent offsets. This provides a role for retailers and brokers (sometimes referred to as aggregators), who acquire offsets from numerous landowners, ensure that offsets are reliably measured and verified, and provide blocks of "clean" offsets to the market. Although retailers and brokers often do not own the offsets, buyers may prefer to work with those that do. That is because such entities are in a better position than individual landowners to self-insure—to make up any shortfalls in the offsets they agree to provide to buyers, who use them to meet their emission caps.

Buyers value certainty in the number of offsets and the time of delivery. Landowners, in contrast, tend to be unwilling to deliver a specified number of offsets at a specified time unless the price of offsets is high, preferring to commit only to certain land-management practices. If a contract requires a landowner to pursue certain land-use practices rather than create a certain quantity of offsets, the buyer assumes the risk that greenhouse benefits will not materialize. Contracts that specify certain land-use practices therefore usually transfer all resulting offsets to the buyer. As an alternative, contracts may specify certain land-use practices and require delivery of a minimum amount of offsets. Another alternative is that, in the future, insurance may become available where the insured could receive compensation if a project does not produce the anticipated amount of offsets.

Accounting for Leakage

Although project developers and quantifiers generally do not have to account for upstream and downstream emissions, they do have to account for *leakage*—emissions displaced from inside the project's boundaries to sources outside it. Leakage is of most concern with projects that aim to boost carbon stocks in existing forestlands by curbing logging. U.S. markets for forest products are robust, and changes in supply in one region often spur compensation in other regions. For example, when logging in the Pacific Northwest was slowed to protect the spotted owl, harvesting shifted to the Southeast. That is leakage on a grand scale.

Landowners aiming to produce offsets by cutting back their timber harvesting must determine how changes in the supply of forest products will affect demand for those products. The landowners must then take the resulting GHG emissions, or loss of carbon stored in biomass, into account in measuring the net greenhouse impact of their project. Assessing leakage can be challenging, but fairly standard economic methods can help (see Chapter 10).

Unlike projects that curb timber harvesting, those that reforest marginal agricultural land are unlikely to result in significant leakage. U.S. farmlands overproduce agricultural commodities, so such projects will probably not spur other farmers to convert forests to croplands to replenish the supply of crops. Such projects will therefore not displace emissions to other lands. Projects that entail a shift from plowing to no-till farming are also unlikely to produce leakage because the latter practice often improves soil fertility and thus boosts crop yields. Therefore, emissions will not move to other locations.

Leakage is a concern if farmers restore marginal croplands in one area to grasslands while converting other grasslands to crops. To address this issue, grassland-restoration projects usually need to establish baseline carbon stocks on a farmer's entire property three to five years before the projects begin.

Unlike uncapped landowners, companies or countries subject to emissions caps do not have to account for leakage if they themselves reduce their emissions; regulators limit total emissions regardless of how they move around among emitters. Consider a situation where one utility reduces its emissions by producing less power, while another utility increases its emissions because it produces more power to replace the lost supply. The second company must still meet its own emissions cap. Thus overall emissions among the capped entities cannot exceed the total established by regulators.

Quantifying Offsets from a Project

Projects will devote significant effort to implementing the protocols for measuring changes in emissions and carbon stocks as detailed in the monitoring and verification plan. To ensure that measurements are both reliable and complete, the crew measuring soil carbon,

biomass, and methane and nitrous oxide emissions in the field must receive specific guidance on how to perform all measurements. Crews should demonstrate proficiency before quantifiers use their measurements to calculate a project's net greenhouse benefits.

Quantifiers use the results of such measurements to calculate the carbon offsets to be claimed by project developers. This process requires five stages.

Stage 1 entails calculating either *C,* the total stock of carbon stored on project lands (in the case of a project based on carbon sequestration), or *E,* the total GHG emissions from project lands (in the case of a project based on reduction of GHG emissions). Quantifiers must adjust *C* or *E* to account for inadvertent emissions arising from the project (see Appendix 2).

Stage 2 entails calculating *B,* the baseline (in the same units as *C* or *E*), and *L,* the leakage.

Stage 3 entails calculating the net GHG benefit of the project, before accounting for leakage. In the case of a carbon sequestration project, the amount of carbon stored on project lands should be greater than the baseline:

$$\text{Net GHG Benefit} = (C - B) \qquad \text{Equation 2.1a}$$

In the case of a project based on reduction of GHG emissions, the emissions from project lands should be lower than the baseline:

$$\text{Net GHG Benefit} = (B - E) \qquad \text{Equation 2.1b}$$

Stage 4 entails calculating the project's carbon offsets by subtracting leakage from the net GHG benefit. By convention, leakage is expressed as a fraction of the total NET GHG Benefit, so

$$\text{Offsets} = \text{Net GHG Benefit} \times (1 - L) \qquad \text{Equation 2.2}$$

where the offsets are expressed as metric tons of CO_2e (see the end of this chapter).

Stage 5 entails accounting for the uncertainty or probability error in the offsets calculated with Equation 2.2. This is critical because as uncertainty rises, regulators of a capped system will generally accept a smaller proportion of offsets as real, and buyers in a voluntary system will generally pay less for the offsets. To assure regulators and buyers that the offsets being marketed are real, the offsets actually credited to the project are typically reduced by the uncertainty in the calculated offsets. In other words,

$$\text{Credited Offsets} = \text{Offsets} - S_{\text{offset, 90\%}} \qquad \text{Equation 2.3}$$

where $S_{\text{offset,CL}}$ is the uncertainty in the offsets at a confidence level of CL% and is calculated using the standard statistical techniques described in Appendix 3. As discussed in that appendix, we recommend that uncertainties be stated at the 90% confidence level (i.e., CL = 90%). This means that there is a one in ten chance that the actual offsets achieved by the project are less than the Credited Offsets calculated from Equation 2.3. Because uncertainty reduces the offsets credited to a project, project developers are confronted with a choice. Projects that rely on less reliable methods will generally entail less operating cost, but they will likely generate larger uncertainty. The challenge for developers is to optimize profits by finding a balance so that the methods used to generate acceptable levels of both uncertainty and operating costs.

Registering and Verifying Offsets

After quantifying offsets, project developers must complete a few more steps before they can easily sell them. To ensure transparency and make verification possible, developers should make public a brief written report that addresses several aspects of the project:

- Where it was located.
- Who was responsible for implementing it.
- What activities landowners or farmers undertook to generate offsets.
- Who was responsible for quantifying and verifying the offsets, and how they did so.
- How many tons of offsets the project generated.
- When the offsets occurred, and how long they will persist.

Concerns among project developers about safeguarding proprietary information should not prevent transparency. Published reports can outline the approach to quantifying greenhouse benefits while providing only general information on land conditions and production techniques. Published information can similarly summarize a landowner's obligations and the

legal instruments used to establish and enforce those obligations rather than include the entire contract.

Project developers then need to enlist an independent party to verify the offsets. Most regulated systems require registration and independent verification. Even though voluntary systems do not, buyers should be wary of any offsets that have not been registered and verified.

Verifiers act as auditors of the offset process. They review the data gathered from project measurements and the methods used to quantify the offsets, to ensure that quantifiers implemented the measurement and verification plan properly and accurately. To avoid conflicts of interest, verifiers should not be involved in the project in any other capacity.

The final step in the process is to register the offsets. This entails filing notice of the offsets with the appropriate agency, which assigns the owner a unique identifying number (just as in the registration of publicly traded securities and land properties). If the offsets are sold, the registration is amended to reflect the transfer. Thus ownership of the offsets is unambiguous, and they cannot be sold to more than one party at a time.

Retiring Offsets

When a company or country buys offsets to comply with its emissions cap, it surrenders those offsets to the regulators of the cap-and-trade system, who "retire" them to ensure that no other company or country can use them. For example, regulators may assign a utility a cap of 100 tons of GHG emissions during a certain period. If the company emits 110 tons, it must buy at least 10 tons of offsets to comply with its cap. When the utility uses the offsets to meet its cap during that period, regulators retire them, precluding anyone else from using the offsets. An environmental group or other concerned party could also voluntarily acquire offsets and retire them by submitting them to regulators.

Offsets that involve sequestering carbon in trees or soil are reversible. For example, timber grown under a forestry project can be cut or burned, re-emitting the stored CO_2. Offsets that entail avoiding GHG emissions, in contrast, are permanent. For example, if a farmer creates offsets through practices that lead to lower emissions of methane or nitrous oxide, even if that farmer later reverts to high-emission activities, the offsets created earlier still continue to exist.

Regulators and buyers must have a mechanism for tracking reversible offsets to ensure that they have not been lost. One approach is to assign them an expiration date. Unless a utility or other buyer submits evidence that the offsets continue to exist by the expiration date, regulators can reclassify them as expired. A company or country that relies on reversible offsets to meet its emissions cap is essentially deferring emissions to the time when those offsets expire. The emitter must then either buy new offsets or show that the offsets remain intact.

Expressing Offsets in Global Warming Potential

Participants in a cap-and-trade system must have a uniform way of quantifying the greenhouse impact of the offsets they buy and sell. That is especially important because different greenhouse gases cause different amounts of warming per unit of gas, and they persist for different lengths of time in the atmosphere.

Global warming potential (GWP) compares the impact of different GHG emissions on the climate over a 100-year period.[3] The GWP of each greenhouse gas is determined by two quantities: the amount of warming a specified amount of the gas causes and the amount of time it persists in the atmosphere. Consider two hypothetical greenhouse gases, A and B. Both remain in the atmosphere for the same length of time, but A causes twice the warming as B. In this case, A would have a GWP twice as large as that of B. Alternatively, imagine that A and B cause the same amount of warming, but A stays in the atmosphere twice as long as B. In that case, A would again have a GWP twice as large as that of B.

By convention, the GWP of CO_2 is assigned a value 1, and the GWPs of other gases are expressed in relation to that of CO_2. Table 1 provides the GWP of the three gases most relevant to land-management practices: CO_2, methane (CH_4), and nitrous oxide (N_2O). Two sets of values are shown, one set from the 1995 report of the Intergovernmental Panel on Climate Change and the other from its 2001 report.[4]

In quantifying offsets, quantifiers use GWPs to convert the tons of CH_4 and N_2O emissions that a project avoids emitting into metric tons of CO_2e (one metric ton equals about 2,205 pounds). For example, in 2001, one ton of CH_4 emissions had the same warming potential as 23 tons of CO_2, and thus it would be 23 tons CO_2e.

Field scientists typically measure the amount of carbon and nitrogen in biomass and soil, rather than the amount of CO_2 and N_2O. Thus, to quantify offsets, analysts need to convert those field measurements into the equivalent amounts of GHGs. For carbon, they do that by dividing the molecular weight of CO_2 (44) by the molecular weight of carbon (12) and multiplying the result (3.67) by the amount of carbon in soil. Thus storing 1 ton of carbon in the soil saves 3.67 tons of CO_2e. Analysts can similarly convert nitrogen to N_2O by dividing the molecular weight of N_2O (44) by the molecular weight of N_2 (28). They multiply the result (1.57) by the amount of nitrogen that becomes N_2O. Thus 1 ton of nitrogen emitted as N_2O equals 1.57 tons of N_2O. Using a GWP of 296 for N_2O, 1.57 tons of N_2O is equivalent to 464 tons of CO_2e. With offsets reliably quantified in tons of CO_2e, landowners can sell them to companies or countries that must meet their emissions caps or to buyers that voluntarily agree to limit their GHG emissions.

This overview of the steps involved in developing carbon offsets based on land-management techniques reveals how challenging that process can be. However, with careful planning and implementation, landowners and farmers can create reliable offsets that help slow global warming while they also reap a new income stream. The next chapter explores the various kinds of offset projects landowners and farmers can pursue. Part II of the book then provides a more detailed discussion of each of the steps outlined in this chapter.

Table 2.1 100-Year Global Warming Potentials

Gas	1995 GWP	2001 GWP
Carbon dioxide	1	1
Methane	21	23
Nitrous oxide	310	296

Source: Intergovernmental Panel on Climate Change (1995, 2001).

Chapter 3

Land-Management Options
for Creating Offsets

Landowners and project developers can pursue a variety of activities to create carbon offsets. This chapter focuses on four overarching types of projects based on those activities:

1. Projects designed to sequester carbon in biomass through the cultivation of new forests and grasslands or through delays in the harvesting of forests.
2. Projects designed to sequester carbon in soils through changes in farming practices, such as through the adoption of no-till farming.
3. Projects designed to reduce greenhouse emissions of methane and nitrous oxide through changes in farming practices.
4. Projects designed to reduce greenhouse emissions of methane through changes in the practices used to process and dispose of manure.

These four types of projects do not encompass every possible change in land management that could occur at every site. (This book does not address activities available to landowners that are not specific and unique, such as building windmills and growing bioenergy crops.) However, they are particularly promising because they can

– Create larger amounts of offsets per acre than other techniques.

– Create offsets at moderate or low cost per ton.
– Create substantial amounts of offsets across the United States.

Although this book addresses each type of project separately, any one change in land-management activity can exert several different impacts on GHG emissions. For example, a project that shifts from cultivating annual crops through plowing and applying nitrogen fertilizer to growing and permanently conserving forest could produce several types of GHG benefits. These include sequestering carbon in trees and other vegetation, sequestering carbon in the soil, reducing fossil-fuel emissions from the machines used in cropping, and curbing nitrous oxide emissions from the fertilizer. Project developers can take all these benefits into account in assessing the amount of carbon offsets their project produces. To do so, however, they must quantify the GHG benefits of changes in each source of emissions separately (see Chapters 6–9). They must also take into account a project's inadvertent emissions, such as from the use of fossil fuels to power tractors, which reduce the amount of offsets a project can claim (see Appendix 2).

Increasing the Amount of Carbon Stored in Trees and Wood Fiber

The most common type of land-management activity for creating offsets—and often the most productive—is to boost the amount of carbon stored in trees and wood fiber. This activity includes four approaches:

- Establishing trees on land that does not have trees.
- Allowing existing trees to grow larger, or otherwise increasing the amount of biomass in forest stands.
- Increasing the amount of carbon stored in wood products and wood waste.
- Decreasing the loss of carbon stored in trees.

Establishing New Trees

Trees are uniquely efficient at removing carbon from the atmosphere and storing it as long-lasting biomass. Oven-dry wood is approximately half carbon by weight—carbon that was all in atmospheric carbon dioxide before the tree converted it during photosynthesis. As it grows, a typical acre of Douglas fir will sequester about 300 to 400 metric tons of CO_2e over its first 50 to 75 years. A healthy 100-year-old stand of Douglas fir on a productive site may hold as much as 800 metric tons of CO_2e.

Practices for establishing trees are well known. On many sites that could support trees but do not, the barrier is financial rather than practical. There are two types of financial barriers. First, other land uses may provide greater financial returns than activities producing GHG offsets. Second, establishing trees usually costs several hundred dollars per acre, and investors may be unwilling to make a substantial cash investment on hope of receiving a financial return on the future sale of offsets. On optimal U.S. sites, the cost of sequestering carbon in trees on former agricultural land can be as low as $6 per ton CO_2e,[1] but rising land prices may make this low sequestration cost unattainable in the future.

Projects that seek to sequester carbon by growing new trees should select suitable lands. A substantial portion of such lands probably supported trees at some point during the last millennium and were cleared for agricultural use. Suitable lands also include those from which merchantable timber has been extracted, but that did not regenerate forest and are not managed. Suitable lands usually do not include native grasslands, and most buyers will not want offsets created by destroying such lands and replacing them with forest. Note that the conversion of native forest to a fiber-oriented plantation may reduce the amount of carbon stored in the forest.

The most cost-effective sites for sequestering carbon in trees are those that are in low-value agricultural use or those with degraded forest that are located where land prices are low because development demand is weak. Such sites may be able to grow trees quickly enough that the revenue from offsets can cover the costs of purchasing the land and establishing trees.

Sites that require irrigation to grow trees are often not good candidates for sequestration projects because of GHG emissions from the fuel used to pump irrigation water and because of the potential for CO_2 emissions when irrigation water chemically reacts with soils and waters. Projects would have to take such emissions into account (see Appendix 2). However, projects that sequester carbon in wood products by irrigating short-rotation trees may be economically feasible, given revenues from both wood products and sequestration.

Allowing Existing Trees to Grow Larger

Project developers can create offsets by increasing the age of trees at final harvest or by ceasing harvest and letting young trees grow. If harvesting ends, greater harvest elsewhere will replace most of the lost supply of wood products. Such a project would therefore have to subtract the emissions from the displaced harvest from the amount of carbon the project sequesters. Projects can mitigate such displacement, or leakage, by including new wood production. For example, a project that preserves forest can include a new plantation that is managed for a high rate of wood production.

Extending timber rotations can sequester a few metric tons of CO_2e per acre each year by raising the average carbon stock on productive timberland. However, because developers of such projects would forego timber revenue, the cost per ton of offsets could be high. Moreover, these projects would not create offsets until the forest reached the longer rotation age, but the costs of foregoing harvest would occur at the start of the project. Still, extending rotation lengths increases not only the total amount of wood at harvest but also the average amount of volume growth per year, partially offsetting the loss from deferring harvest. At today's wood prices, offsets created by extending rotations would usually cost more than $100 per ton. However, they could cost less if the species or quality of the trees is low or if the landowner has some other reason for deferring harvest.

In theory, another strategy for increasing forest carbon stocks is reducing carbon losses from disturbances such as fires and disease. However, in practice, such disturbances are hard to control. In addition, efforts to reduce disturbances (such as vigorously thinning overstocked stands to curtail the risk of fire and disease) may also produce more GHG emissions from the added use of fossil fuel than the sequestered carbon offsets.

Increasing the Carbon in Wood Products and Wood Waste

Wood products such as paper and lumber hold carbon sequestered from the atmosphere. Projects may increase such sequestration by boosting the proportion of harvested wood that forest owners set apart for wood products.

Whether carbon sequestered in wood products counts as offsets depends on the GHG mitigation system. A regulatory or voluntary system may assign ownership of such offsets to tree growers, wood product manufacturers, whoever possesses the wood fiber, the government, or someone else. If the system assigns ownership to the possessors of wood fiber, very few will have enough carbon storage to justify the costs of documenting it, so most such sequestration would go uncounted. If the system assigns ownership to landowners who grow trees, the offsets could provide a noticeable increase in the financial returns from forestry, thus encouraging the growth of trees for wood products. If the system assigns ownership to tree growers or product manufacturers, the amount of carbon stored is a function of the lifespan of various products, based on studies of how long they last.

With efficient timber harvest, only about a quarter to a third of the carbon in live trees usually ends up in products. Whereas entire tree trunks may be used to make paper or other wood products, this stem wood contains only a bit more than half of the carbon in a merchantable tree. Processing the stem wood produces mill waste, which further reduces the carbon from trees that goes into wood products. Some of the mill waste is used for products such as particle board (which could be counted as sequestered carbon for the life of the board), but some is burned for heat or electricity (which could not be counted as sequestered carbon, but which might avoid fossil-fuel emissions) Nevertheless, the rate of carbon flowing into products can be significant. For example, tree growth on a productive site hosting Douglas fir in the Pacific Northwest can average more than 900 board feet per acre per year. This represents slightly more than 1 metric ton of CO_2e of carbon in lumber.

Carbon sequestration in wood products can rise even when carbon stocks within a forest are constant. In a forest where harvest equals growth, and the amount of carbon in woody debris and on the forest floor remains constant, the total amount of carbon in the forest will not change appreciably. Yet wood is flowing out of the forest, and any wood products will sequester carbon. Of course, wood products are taken out of use over time, so project developers and buyers must apply decay rates to offsets based on sequestration in these products.

The ultimate fate of the wood products also affects the amount of offsets that a project can create. When buried in an engineered and managed landfill, most carbon in wood waste remains stored almost indefinitely. Buried waste usually releases a small amount of methane, but this usually cancels out only a portion of the sequestration. However, if waste is burned or left to decompose, it releases essentially all its carbon into the atmosphere, reversing all the initial offsets awarded to a project for sequestration in wood products. If the waste decomposes in anaerobic conditions, it could release enough methane to lead to a negative offset.

GHG mitigation systems may allow project developers to count carbon stored only in wood products that remain in use, or they may also allow them to count carbon in products disposed of in landfills. If the latter is the case, either the landfill operator or the project developer can claim credit for the offsets depending upon who owns the rights to the carbon stored in the landfill.

Decreasing the Loss of Carbon Stored in Trees

Carbon released from forests accounts for a significant proportion of anthropogenic emissions.[2] In the United States, forest is usually cleared for residential development.[3] In developing countries, forests are usually cleared for agriculture. Reducing those losses would not only provide greenhouse benefits but also help maintain the ecological value of forests.

However, simply preventing the removal of trees from a specific area, without addressing the reasons why removal is occurring, is unlikely to result in much net GHG benefit. Most demand for wood or land will be displaced to other locations.

Project developers may satisfy demand for wood by establishing a highly productive forest plantation on land not previously forested, usually land in agricultural use. Displaced demand for agricultural land could be offset through lower agricultural subsidies or the use of more productive or sustainable farming methods. Incentives for clustered development, or regulatory limits on development, could also reduce demand for land. For example, a landowner could voluntarily cluster development—indeed, clustering is occurring in many forested rural regions of the United States to preserve habitat—or a zoning board could require clustering. Any resulting offsets would have to undergo an additionality test and account for leakage. However, they would be legally strong if both the landowner and the zoning board acknowledge the potential claims of the other and share the benefits.

If trees have value as timber, and the landowner requires payment of at least the value of the timber to preserve the trees, offsets are likely to cost at least $70 per metric ton CO_2e in the United States. The cost of sequestration can be much lower if there is no alternative economic use of the wood fiber. The cost can be much higher if a project must buy all the timber on a property each year but can count only a small portion of the total carbon stock as an offset each year because clearing affects only a small proportion of the forested area.

This option is financially viable only if offset prices are relatively high.

Increasing the Amount of Carbon Stored in Soil

Globally, soil and plant detritus contain 1.5 to 2 trillion metric tons of carbon—nearly three times the amount in the atmosphere. Yet conventional agricultural practices, especially plowing, have reduced soil carbon stocks. This situation has created an opportunity for landowners: by replenishing these lost carbon stocks, they can remove carbon dioxide from the atmosphere and can market the stored carbon as offsets. Practices that restore carbon also improve food security because they help soil retain moisture and nutrients while making those nutrients accessible to growing plants. That, in turn, reduces the need for nitrogen fertilizer, thereby cutting nitrous oxide emissions.

Over time, the amount of carbon in soil moves toward equilibrium, when carbon removals exactly balance carbon inputs. Land-management practices that change the rate of carbon input, rate of carbon removal, or both can raise the equilibrium level and increase carbon stocks.

Strategies that can increase soil carbon include:

– Curbing soil disturbance from tillage when growing annual crops.
– Increasing carbon inputs from plant residue by boosting the rate of plant growth or the proportion of time that plants are growing and by leaving the biomass onsite.
– Increasing the proportion of plant biomass retained onsite.
– Switching to perennial species such as grasses and improving grassland conditions.
– Changing conditions to favor, to a degree, the formation of inorganic carbon compounds.

Decreasing Soil Disturbance in Annual Cropping

In conventional tillage, plowing completely inverts the soil after the harvest of one crop and before the planting of the next crop. The aeration and exposure of organic matter spurs activity among microbes that use organic materials as food, thereby decomposing it and releasing CO_2 into the atmosphere.[4] Tillage provides immediate benefits by releasing some nutrients previously protected within soil aggregates, controlling weeds, and speeding soil warming in the spring. However, over time, these benefits come at the cost of degraded soil quality.

Conservation tillage encompasses a range of methods for preparing the seedbed that curb erosion and leave at least 15 to 30 percent of the surface covered with crop residue immediately after planting (the specific percentage depends on the crop). Conservation tillage also requires fewer passes of equipment than conventional tillage and thus reduces fossil fuel emissions.

Such methods include no-till and reduced-till cropping, which disturb the soil less than conventional tillage, and other practices that leave substantial amounts of crop residue on the soil surface. Conservation tillage raises the proportion of organic carbon in the soil by avoiding increases in soil respiration, which converts organic carbon to CO_2 released into the atmosphere. Reducing tillage also allows clumps of soil to retain carbon-containing organic material, preventing it from decomposing.

Farmers using *no-till cropping* do not disturb the soil except to carve slots in the ground for seed. No-till farmers also leave crop residues on the ground, which, along with herbicides, help control weeds. Farmers practicing *reduced-till farming* (also called strip plowing) plow only narrow strips of soil, breaking up the soil less than conventional plowing. Reduced-till practices include ridge tilling, wherein farmers make ridges every two or three years and grow a row of crops on each ridge, leaving the soil undisturbed between harvest and planting.

Another type of conservation tillage is *chisel plowing*, wherein farmers use vertical shafts to rip the soil, aerating it but leaving some crop residue to protect the surface from erosion. Chisel plowing leaves a rough surface and requires disking to smooth it.

Not all crops are amenable to no-till farming. Removing root crops, such as potatoes, from the ground disturbs the soil, and farmers sometimes need tillage to control pests or diseases. For example, California regulations require farmers to shred the root crowns of cotton plants every fall to control pink boll worm, which they do using tillage.

Recent analyses have shown that reduced-till practices sequester carbon only under certain climatic conditions. Regions with relatively high precipitation (such as the U.S. Corn Belt, which extends from Indiana to Iowa) have a fair potential for storing carbon because they have high productivity, which provides large amounts of crop residue as carbon input to the soil. Drier environments (such as the Great Plains, west of the Corn Belt) may not generate enough sequestration to make offset projects financially viable. That is because lower rates of productivity yield only modest carbon inputs, and soil respiration over a larger proportion of the year may actually reduce carbon input to the soil.

Reduced-till practices such as chisel plowing can still disturb the soil enough that it stores little carbon. Unlike reduced-till farming, no-till cropping can restore some of the carbon lost through tillage. Although sequestration rates vary with climate and crop rotation, no-till farming boosts carbon storage in moist, semiarid, and arid conditions alike because it slows respiration enough to raise the equilibrium soil carbon content. Increases in soil carbon are greatest when farmers couple a switch to no-till cropping with intensified cropping (see the next section). However, no-till farming does not expand carbon stocks in cold, moist climates where soil carbon levels are already relatively high, such as in parts of eastern Canada.

Rotational-till systems use both conventional and conservation tillage practices, depending on the crop, to control pests and weeds. However, rotational systems are not particularly effective in sequestering carbon. Studies have shown that carbon storage declines dramatically if land under continuous no-till management reverts to conventional tillage. Even a single pass of the plow every four to five years releases much of the carbon stored through previous no-till practices.

Switching from plowing to no-till farming may increase nitrous oxide and methane emissions as soil structure recovers from plowing. The mechanisms that cause these increases are not completely understood, but they probably reflect low oxygen levels within clumps of soil that form after tillage ends. The increases in methane and nitrous oxide emissions can cancel out the GHG benefits of carbon sequestration resulting from no-till farming for several years. Because most of the extra GHG impact stems from nitrous oxide emissions, developers of no-till projects must carefully control the timing and amount of nitrogen fertilizer they apply. They must also continue their projects for at least a decade to counteract this effect. Also, in many parts of the country, a substantial portion of cropping already uses conservation tillage. As a result, the baseline for a project will reflect this existing proportion of conservation tillage, and a proportion of the carbon sequestration achieved by the project equal to the proportion of existing conservation tillage will not count as an offset.

Increasing Carbon Inputs from Crop Residue

Plant residues converted into organic matter are the major source of carbon in soil, so leaving plant residues behind after harvesting is an important technique for enhancing soil carbon. Farmers can raise the amount of crop residue by accelerating the growth of existing crop strains or switching to crops that produce more biomass. For example, soybeans produce relatively little residue, whereas corn generates large amounts. If farmers leave crop residues on fields, switching from soybeans to corn will increase carbon inputs to the soil, all other aspects being equal. The extent to which such activities can lead to carbon offsets depends on a variety of factors that landowners need to carefully consider before they embark on an offset project.

Most crop residues decompose over a few years. A fraction of 1 percent of such residues typically becomes humic material, which can persist in the soil for centuries. Decomposition converts most of the remaining carbon to CO_2, which is returned to the atmosphere. The persistence of soil organic carbon is inversely related to temperature: cold soils can build huge stocks of organic matter, whereas many tropical soils never accumulate much, regardless of inputs and land-management techniques.

If farmers expand residue inputs for only a few years and then return to lower levels, much of the gain in soil carbon will quickly be lost, as microbes consume easily decomposed plant material and release CO_2. However, if farmers of moderately productive, temperate soils sustain high residue levels for a decade or two, a significant proportion of the gain in soil carbon from humic materials should persist long after the farmers revert to low input levels.

Mineral fertilizers and better crop varieties raised crop yields steadily during the latter half of the twentieth century. This increased the amount of plant residue, and thus carbon, in the soil.[5] In the U.S. Corn Belt, rising amounts of crop residues partially reversed losses of soil carbon from tillage in the first half of the last century. Further increases that may even surpass natural soil carbon levels may be possible through the application of organic fertilizers such as manure and compost.

Raising levels of organic carbon in soil can increase nitrous oxide emissions, which have a greater warming impact than the carbon sequestration mitigates. This is most likely to occur when projects start with very low levels of soil carbon and when nitrogen is plentiful in the soil. Moreover, if farmers obtain manure by adding less of it to other locations, then the carbon gains may come at the expense of carbon losses from the soil at other sites.

Enhancing crop production can exert negative as well as positive GHG impacts on-site, upstream in the production chain, or downstream. The manufacture of chemical fertilizers typically causes substantial upstream emissions. In some cases, sequestration of carbon in soil can more than offset these emissions, but project developers must analyze the net impact. Under continuous tillage, the CO_2e of soil carbon sequestration resulting from increased carbon inputs to the soil from increased fertilization appears to often be less than the CO_2 emissions from manufacturing the additional fertilizer (Schlesinger 2000). However, land managers may also apply more fertilizer than plants and soil biota can absorb. This excess nitrogen can cause significant nitrous oxide emissions, and it is the major source of downstream water pollution in some regions.

Irrigation enhances crop production and therefore soil carbon, particularly in semiarid and arid regions. However, increasing irrigation typically raises emissions of methane, nitrous oxide, and CO_2, canceling out some or all of the carbon sequestration. If efforts to expand crop production rely on more irrigation, project developers should account for any resulting methane, nitrous oxide, and CO_2 emissions, and for emissions from the fuel used for pumping. Rising amounts of crop residue can also sometimes lead to methane and nitrous oxide emissions. Although both dry and moist soils are generally methane sinks, residue that decomposes while saturated with water produces methane. Farmers may also burn residues to clear the soil surface for later crops, control disease, or release nutrients. If combustion is incomplete, this burning produces methane. Again, project developers must take these emissions into account.

Boosting crop residue by increasing the proportion of time that plants are growing on a site usually entails eliminating or reducing fallow periods, when the soil is bare. For example, if farming focuses on a summer crop, and a field has lain bare between fall harvest and spring planting, cover crops such as rye and legumes enhance carbon storage because of the added residues they produce. Farmers often treat cover crops as green manure, leaving the whole plant in the field to provide maximum carbon input. Although increasing the proportion of time that plants are growing enhances carbon storage, coupling more growing time with no-till farming is much more effective because this both increases carbon inputs and decreases removals.

If farmers were to allocate the cost of cover crops only to carbon sequestration, the cost per ton would be moderately high. However, the cost per ton of sequestration could be modest if the cover crop is a nitrogen fixer, reducing the need for chemical fertilizer, and if it enhances soil quality and thus crop production. Switching from chemical fertilizers to cover-crop residues may also cut nitrous oxide emissions, much like switching from chemical fertilizer to manure.

Bare-summer fallow has proven an important land-management option in semiarid and arid regions. Water infiltrating the soil accumulates during the fallow period, when crops are not actively growing and transpiring. Greater water availability enhances production during the next growing season, which is then followed by another fallow period. Unfortunately, soil respiration continues to remove carbon from the soil during the fallow period. Under a no-till system, residues limit water lost through evaporation, enhancing carbon storage in semiarid and arid soils. Farmers can crop continuously on some sites using no-till practices, whereas they might need fallow periods if they were using conventional tillage.

Switching from Annual Crops to Perennial Plants

Over time, croplands converted to perennial cover such as grass or some kinds of trees approach native conditions in the amount of organic carbon they store. The switch lessens soil disturbance and improves soil structure. Although the rate of carbon storage is considerably slower than the carbon loss that occurred after the land's original conversion to cropland, the soil is likely to gain more carbon than under other management options.

For example, the U.S. Conservation Reserve Program, which encourages landowners to set aside marginal cropland for grassland and forest, has sequestered more carbon in U.S. agricultural lands than any other program. As of 2004, the program had enrolled nearly 35 million acres with grassy or woody perennial vegetation.[6] However, the total sequestration is small relative to total emissions from agriculture.

When quantifying the greenhouse benefits of converting cropland to grassland or forest, developers must estimate leakage—the other lands planted in crops as a result of the loss of project cropland. Research shows that about 20 percent of a conserved area is canceled by other lands brought into production elsewhere (Wu 2000). The amount of leakage may be more or less than the proportion of area displaced (Murray 2004).

Mixed systems incorporate one to several years of pasture or hay into a crop rotation, such as four years of corn followed by four years of hay. Stands of perennial grasses such as hay tend to have more root mass than annual plants, so switching from annual crops to a mixed system can increase carbon stocks by expanding below-ground residue. However, the gains will not be as great as under a permanent switch from annual crops to perennial species, particularly if farmers use tillage to clear perennial plants for an annual crop.

Some projects may consider harvesting plant residues for use as fuel. Developers should analyze such projects carefully because removing residues reduces carbon inputs into the soil. The net effect of projects that remove aboveground biomass from a site will depend on the soil carbon stock, the amount of residue left on the soil, and the speed at which the retained material decomposes. Even so, if residues displace fossil fuel used for energy, this mitigation will usually be much greater than the amount of soil carbon lost.

Improving Grassland Conditions

Several grassland-management practices influence the carbon balance in soils, and thus they could provide the basis for offset projects. These practices include reducing grazing intensity, changing the timing of grazing, irrigating grasslands, applying mineral and organic fertilizers, seeding more productive varieties and legumes, and reducing or eliminating burning. The more landowners use these practices in managing pastures and rangelands, the more carbon the lands will store.

Adding water or nutrients can increase plant growth and thus the amount of plant biomass available for carbon storage. Grazing has an impact on carbon storage because it removes biomass, but the effects are variable because some plant species respond to some types of grazing by allocating more carbon to roots than to shoots. Under many conditions, a higher proportion of root residues than aboveground biomass residues are incorporated into soil organic matter. As a result, some grazing regimes that improve grassland conditions can sequester carbon. When the grazing regime does not have this effect and leaves more carbon in the shoots, biomass must be transferred to the soil through physical mixing such as earthworm activity or through leaching of dissolved organic carbon into the soil.[7]

The net impact of fire on the storage of organic carbon depends on two contradictory effects. Intuitively, burning surface organic matter would seem to limit the potential for increasing soil carbon because that practice converts plant matter into atmospheric CO_2. However, burning stimulates new plant growth, in part by releasing nutrients bound in plant tissues. In moist climates (such as the eastern prairies of the United States),

plant growth from nutrients released by fire offsets the removal of residue, leading to a minimal loss or even an increase in carbon storage. A portion of the burned material also turns into charcoal, which can persist for millennia in the soil (although if sites burn again, the charcoal may also burn).

The increase in plant production in response to fire depends on several factors, including climatic variability, topography, and grazing. The net effect of fire on carbon storage is therefore highly variable. These impacts range from negative under conditions such as drought, which limits later plant production, to positive in situations where regrowth (with or without charcoal production) outweighs carbon consumed by fire.

Sequestering Inorganic Carbon

Most soil carbon projects target organic carbon, which is derived from the tissues of plants or animals and is contained in organic molecules. Inorganic carbon, though, is an important component of total soil carbon levels, particularly in arid soils. Inorganic carbon can be very stable and thus remain in soil indefinitely. However, potential rates of inorganic carbon sequestration are low, and in most soils, they are less than potential sequestration rates for organic carbon. Irrigation can cause noticeable rates of deposition of inorganic carbon in soils, but the effect of increasing water can cause substantial amounts of net CO_2 emissions from soils. Pumping irrigation water also causes substantial CO_2 emissions if fossil fuel is used as the source of energy.

Soils can either produce or destroy inorganic carbon, depending on several factors, but mainly on soil pH. Alkaline soils usually convert CO_2 from the air into solid carbonate, especially calcium carbonate. Acidic soils usually break down carbonates and bicarbonates in the soil, releasing the carbon in these molecules as CO_2. Irrigation of soils with water containing substantial amounts of calcium bicarbonate can release inorganic carbon as CO_2 because of acidification resulting from greater plant growth and nitrogen fertilizer. If a project entails irrigating arid soils, it should assess emissions from inorganic soil carbon.

Cutting the amount of fossil fuel used to manage lands reduces greenhouse gas emissions directly. Land managers can curb fuel use by switching to cropping practices that require fewer passes of equipment across fields or to practices that require less power for each pass. For example, plowing requires more power than most other cropping activities.

Eliminating plowing by switching to no-till farming can save a significant amount of fuel while requiring only modest amounts of greater effort during planting and spraying. Typically, projects that switch from conventional plowing to no-till farming save about 2 gallons of fuel per acre per year. Every gallon of fuel burned emits about .01 tons of CO_2. For a farm that cultivates 2,000 acres, reducing emissions by 2 gallons per acre per year would cut total emissions by about 40 metric tons of CO_2e per year. If land managers switched back to conventional tillage, they would stop generating new offsets from cuts in fuel use, but they would not reverse offsets from previous years.

Land managers can also reduce emissions by switching from fossil fuel to biofuels. However, any net reductions in greenhouse gases must account for emissions from the production of the biofuels, which can be significant. Verifiers must also ensure that cuts in emissions are counted only once, rather than by both the fuel manufacturer and the end user. (This book does not include guidelines for calculating offsets from biofuels. The fuel sector has developed those methods, and they do not relate directly to land-management practices.)

Reducing the Use of Nitrogen Fertilizer

Biological and geochemical processes in soil can convert nitrogen from fertilizer to nitrous oxide, which is emitted to the atmosphere. Because one pound of nitrous oxide has about 300 times the warming effect of one pound of CO_2, a relatively small amount of nitrous oxide emissions can offset the entire greenhouse benefit of sequestering carbon in soil.

Land-management practices, especially agricultural practices, greatly affect the amount and transformation of nitrogen in soils. The major human sources of nitrogen in soils occur when farmers apply fertilizers and manure and when they incorporate crop residue into soil.

To mitigate nitrous oxide emissions, landowners can either reduce the amount of nitrogen in the soil or interrupt nitrification and denitrification. Nitrification is an oxidation process that typically releases energy and produces nitrate. Under some conditions, nitrification can result in some of the nitrogen leaking out as nitrous oxide. Denitrification entails a chain of microbiological reactions that mostly produces gaseous nitrogen (N_2) but can also emit nitrous oxide or nitric oxide from intermediate steps in the process.

The primary option for reducing nitrous oxide emissions from soil is curbing the use of nitrogen fertilizer. U.S. farmers often apply more nitrogen in fertilizer than crops can absorb. Testing soil to determine its nitrogen content, calculating the amount the next crop needs, and adding just enough to counter any shortfall may allow land managers to reduce their fertilizer use, thereby saving money as well as reducing nitrous oxide emissions. Such practices can also reduce water pollution because the excess nitrogen that crops and microbes cannot absorb can leach into streams and groundwater during wet periods.

Timing nitrogen inputs to match crop demand for nitrogen is another strategy for reducing nitrous oxide emissions. The greatest need for nitrogen occurs when crops are adding biomass and forming seeds, which usually have a higher nitrogen content than other plant tissues. This growth generally occurs in the spring and, in regions with summer moisture, summer. However, farmers often apply fertilizer in the fall because the soil is firm and can support the weight of heavy equipment and because they often have more time after harvest. Fertilizing during planting or after crops emerge (often by using lighter existing tracks or large tires to avoid soil compaction) can decrease nitrous oxide emissions and also translate into a direct financial benefit. Nitrification inhibitors, which slow the release of nitrogen, can further improve the match between the timing of nitrogen release and crop demand for nitrogen.

Ammonia and ammonium bicarbonate usually have higher nitrous oxide emissions per unit of applied nitrogen than other common forms of fertilizer. Thus farmers can reduce nitrous oxide emissions by switching from ammonia or ammonium bicarbonate to another form of fertilizer.

Soils that are very acidic—that is, those with a pH of 5 or lower—produce more nitrous oxide than other soils. Land managers may want to consider liming such soils to decrease their acidity. However, liming also produces CO_2 emissions, and developers should consider those emissions when assessing a project's viability.

If the entire life cycle of agricultural production is considered, the CO_2 emissions that result from manufacturing nitrogen fertilizer are large. However, if a farmer reduces use of nitrogen fertilizer, the fertilizer manufacturer does not necessarily make less fertilizer, and emissions may not go down. Even if the emissions from manufacturing fertilizer decrease, this decrease will show up in the emissions inventory of the fertilizer manufacturer. The farmer who uses less fertilizer does not get to count the emission reduction of the fertilizer manufacturer unless the farmer and the fertilizer manufacturer have an agreement that the farmer will count the emission reduction and that the fertilizer manufacturer will not count that reduction.

Reducing the Frequency and Duration of Flooding

Although the processes governing methane and nitrous oxide emissions from soil are complex, those emissions result mostly from just a few steps. Both processes require that soils be depleted of oxygen, which most often occurs in water-saturated soils. In methane generation (or methanogenesis), soil microbes use carbon compounds to produce energy under low-oxygen conditions, with methane as one output. Channels in the soil or plants can speedily transport methane from deep in the soil to the atmosphere. However, if the methane diffuses into high-oxygen microsites within the soil, oxidation can consume the methane before it is emitted to the atmosphere. The balance between the production and oxidation of methane determines the net flow between the soil and the atmosphere. Nitrous oxide production occurs during denitrification. In this process, soil microbes use carbon compounds to generate energy while consuming soil nitrate. This

process produces molecular nitrogen (N_2) and nitrous oxide (N_2O), which diffuse through the soil and into the atmosphere.

Methanogenesis and denitrification occur under different environmental conditions, but they share two characteristics. Both result from the activities of soil microbes, as affected by temperature, moisture, pH, and other environmental factors. In turn, a few ecological conditions—climate, topography, soil properties, vegetation, and human activity—govern those environmental factors. A change in the ecological conditions can change the environmental factors and thus how much nitrous oxide and methane the soil emits. For example, intensive grazing can compact soil and limit water and air flows, creating anaerobic conditions (which increases methane production) or prolonging dry soil conditions (which increases methane consumption).

A large proportion of methane emissions from soil stem from rice paddies and wetlands. Changing plant species or rice cultivars can reduce the amount of carbon exuded by roots, which reduces the input available for methanogenesis, significantly curbing methane emissions. The transport of methane from wetland soils to the atmosphere occurs mainly through plant stems and roots, particularly through gas-filled tubes within those components. Land managers can reduce the transport and release of soil gases by replacing plant species or cultivars with others that do not have internal tubes that effectively transport gas.[8] For rice growers, the greatest opportunity for reducing methane emissions entails switching from rice varieties that need flooding to upland varieties that are grown without flooding.

Reducing Emissions from Anaerobic Decomposition of Waste

Landowners can often produce offsets by switching from a manure-handling system that emits high levels of GHGs to a system that emits low levels. Storing manure with limited aeration usually results in anaerobic decomposition, which produces methane. In contrast, treating manure so it is thoroughly aerated allows aerobic decomposition, which generates very low emissions. Conversion of manure to methane ranges from nearly complete in anaerobic digesters to nearly nothing in intensive aerobic digesters and daily spreading of residue on fields. A practice that produces large amounts of methane can have low emissions if it captures and destroys methane before it is released into the atmosphere.

The Intergovernmental Panel on Climate Change has defined common manure-management practices with higher methane production (from highest to lowest emissions):

Anaerobic lagoon: Waste is flushed with water to open ponds, where it is stored for more than a month, with or without the capture of methane or the use of water that remains after solids settle.

Anaerobic digester: Solids are converted to methane with the help of a slurry of dung and urine, depending on temperature control, mixing, or pH management. The resulting gas may be released, flared, or used to generate power.

Liquid slurry: Dung and urine are transported and stored for months in liquid form, with water added as needed for handling, in tanks open to the atmosphere.

Pit storage: Combined dung and urine are stored in vented pits below stalls.

Deep litter (cattle and swine): Dung and urine accumulate in stalls for long periods.

Common manure-management practices with low methane production (from highest to lowest emissions) include the following:

- *Dry lot*: In dry climates, litter is allowed to dry in stalls before it is removed.
- *Pasture*: Waste from pastured or range animals is left where deposited and not managed.
- *Poultry manure*: Waste is collected in cages, with or without bedding.
- *Solid storage*: Dung and urine are collected from stalls and stored for months, with or without drainage of liquid; this is followed by another use or disposal method.
- *Composting—extensive*: Waste is collected, piled, and turned regularly for aeration.
- *Daily spreading*: Waste is collected daily from barns and spread on fields.

- *Composting—intensive*: Waste is placed in a vessel or tunnel with forced aeration.
- *Aerobic treatment*: Waste is collected as liquid and managed with forced aeration to allow nitrification (the conversion of ammonium to nitrate) and denitrification.[9]

Treatment produces large amounts of sludge, with the extent of methane emissions depending on how the sludge is managed.[10]

Table 3.1 shows the typical methane emissions from each practice. (The warming potential of nitrous oxide emissions from manure is usually small relative to that of CO_2 and methane emissions.)

As mentioned earlier, landowners can take two approaches to reducing greenhouse gas emissions from manure. One is to switch from a high-emissions process to a low-emissions process, such as from an anaerobic system to daily spreading. The other approach is to capture methane from a high-emissions system and burn it. When methane is burned, it produces CO_2, which is a GHG with a lower warming effect than methane. Although projects that capture and burn methane may raise the percentage of manure converted to methane, burning destroys the methane and thus prevents it from acting as a GHG. However, the collection system must be efficient; if more than a tiny proportion of the methane leaks into the atmosphere, net emissions may be greater than those from a low-emissions system.

When methane is burned and converted to CO_2, emissions are the same as if the manure had decomposed. The carbon in methane comes from animal feed, which is usually plant material grown the year before the animal ate the feed. Because plants grow by capturing atmospheric CO_2, burning the methane from manure simply returns that carbon to the atmosphere. Thus, the net offset is the difference between the warming effect of the destroyed methane and the warming effect of the resulting CO_2.

Growing pressure to control odors and air pollution, combined with new rules limiting air and water pollution, have spurred interest in lower-emissions systems for handling liquid manure. In particular, stricter limits on water and air pollution have prompted operators of confined animal feeding operations (CAFOs) to

Table 3.1 Methane Production from Different Systems for Managing Manure under Different Temperatures

System	Percent of potential CH_4 production achieved, cool	Percent of potential CH_4 production achieved, temperate	Percent of potential CH_4 production achieved, warm	Notes
Pasture	0.01	0.015	0.02	
Daily spread	0.001	0.005	0.01	
Deep litter	0	0	0.3	Cattle & swine; storage < 1 month
Deep litter	0.39	0.45	0.72	Storage > 1 month
Poultry	0.015	0.015	0.015	
Solid storage	0.01	0.015	0.02	
Dry lot	0.01	0.015	0.05	
Pit	0	0	0.3	Storage < 1 month
Pit	0.39	0.45	0.72	Storage > 1 month
Liquid slurry	0.39	0.45	0.72	
Anaerobic lagoon	0–1	0–1	0–1	Depends on rate of CH_4 capture & destruction
Anaerobic digester	0–1	0–1	0–1	Depends on rate of CH_4 capture & destruction
Composting-extensive	0.005	0.01	0.015	
Composting-intensive	0.005	0.005	0.005	
Aerobic	0.001	0.001	0.001	

Notes: Cool is defined as an average annual temperature less than 15°C; warm is defined as an average annual temperature greater than 25°C; and temperate is defined as an average annual temperature between warm and cool. CH_4 production is the proportion of carbon input that is converted to CH_4.
Source: IPCC 2000.

adopt new ways of treating large, concentrated amounts of manure. A single operation that houses over a thousand cows or tens of thousands of chickens can produce tens of thousands of tons of manure—and significant quantities of GHG emissions—each year.

Feed for confined animals is usually grown far from the CAFO, and animals that eat shipped-in feed usually produce far more manure than nearby lands can absorb. Farm managers often wish to avoid the signifi-

cant transportation costs of returning untreated manure to the fields that grew the feed. Anaerobic digestion of manure dramatically reduces the volume of material that must be transported. Digesters can also reduce odors and water and air pollution in lagoons used to store manure.

Anaerobic digesters maximize methane production by storing and processing liquid manure under low-oxygen conditions and by providing warm condi-

Table 3.2 The Potential of Land-Management Options to Create Offsets

Activity	Tons per acre over time	Cost per ton	Total U.S. supply of offsets at moderate cost
Increasing carbon stored in existing forests	Moderate	High	Low
Decreasing carbon lost from forests	High	Moderate	Low?
Afforestation	High	Low	High
Increasing soil carbon	Low	Low to moderate	Low
Reducing fuel use	Low	Low	Low? Moderate?
Reducing emissions from nitrogen fertilizer	Low	Low	Low? Moderate?
Reducing flooding	High?	Low?	Moderate?
Changing waste management	High on some sites	Low to moderate	Moderate? High?

Notes: *Afforestation* entails establishing trees on lands where they do not currently exist. *Reforestation*, in contrast, refers to reestablishing trees on sites where they were recently removed by harvest or another disturbance. No specific length of time separates afforestation and reforestation. However, under the Kyoto Protocol, afforestation refers to lands that have not been forested for at least 50 years. Eligibility for GHG credits from reforestation is limited to lands that did not have trees as of December 31, 1989.

tions that speed the metabolic activity of bacteria that digest manure. However, rather than releasing this methane to the atmosphere, farmers can capture and burn it, converting the methane to CO_2 and potentially producing offsets. (A properly managed anaerobic digester can also avoid releasing ammonia, which becomes an air pollutant in high concentrations.) This approach is particularly promising for producing offsets because use of such digesters is not yet the standard approach to manure management. The methane produced by a digester can be burned in a flare, or it can be cleaned, dried, and burned in a generator to produce electricity.

As the cost of digesters falls, prices for electricity and offsets rise, and limits on air and water pollution become more stringent, more U.S. farmers may gain by converting wet, open systems for handling manure (such as lagoons) to closed anaerobic digesters. Such a move can reduce emissions by up to 5 metric tons of

CO_2e per dairy cow each year and by up to 0.4 metric tons of CO_2e per pig each year.

The Bottom Line

The benefits of GHG mitigation projects must be greater than the costs. If project developers intend to create offsets for sale, that means revenues must be higher than costs, adjusted for risk.

Boosting revenues requires producing large amounts of offsets at low or moderate cost. Modest changes in land use do not usually produce large amounts of offsets. Some land-use changes can provide moderate amounts of offsets, but land managers are likely to consider such changes significant shifts in how they do business, and thus they may be reluctant to pursue them (see Table 3.2).

Converting open land to forest can generate several hundred tons of CO_2e of offsets per acre, usually over several decades. Switching to contained manure can reduce GHG emissions by thousands of tons per farm per year. Yet afforestation and anaerobic digesters require significant up-front capital investments. Activities that produce offsets at very low cost may include reducing fuel use and changing the use of nitrogen fertilizer. Land managers may also take inexpensive steps to increase soil carbon, though they may achieve annual offsets of only a ton of CO_2e or less per acre. Maintaining those offsets precludes later plowing or land development. Landowners may regard these constraints on future options as a significant cost, even though the cash cost of sequestering the carbon may be low or negative.

For offset projects to be financially viable, they must usually be compatible with the creation of the products that yielded the bulk of preproject revenues. Project developers must carefully examine their production processes to see if a particular change in activities is compatible with continuing to produce revenue from their lands.

Part II **Steps in Determining a Project's Offsets**

Chapter 4

Step 1: Scoping the Costs and Benefits of a Proposed Project

Before committing to a project, developers will want reasonable assurance that it can provide enough offsets to justify the probable costs. To develop that assurance, developers must "scope" the project.

Scoping requires following the same basic steps as those used to actually quantify a project's offsets. However, the procedures are less rigorous in the scoping phase and therefore far less resource intensive. These steps include:

- Assessing additionality and the baseline.
- Estimating the likely changes in GHG emissions and carbon stocks on project lands, including inadvertent emissions from the project's land-management activities.
- Calculating the net GHG benefits from the project (the difference between the project's changes in both GHG emissions and carbon stocks and the baseline).
- Estimating leakage from the project.
- Estimating the GHG offsets that the project can produce, expressed as tons of CO_2e.

The scoping process should also assess the uncertainty in the amount of offsets estimated for the project (see Appendix 3). If this uncertainty is large, developers face considerable risk that the project will not produce the expected offsets. Risk is one element that project developers will want to consider when deciding whether to proceed.

Developers also care about the timing of offsets, so scoping should estimate the tons a project will produce in each accounting period throughout its life. For example, if demand for offsets in the next year is likely to be large because potential buyers must meet specific emissions caps, a project that delivers offsets in 10 years may not be satisfactory regardless of the offsets' price or certainty.

If initial estimates show that a project is likely to create GHG offsets, developers can then estimate the project's costs and net financial return. If these analyses reveal barriers (such as that one component of the project is very costly or has a high risk of failing), developers may revise the project's design and then scope it again. Developers may also use scoping to compare multiple project scenarios and to consider potential shortfalls or windfalls in the projected amount of offsets. Taken together, this information enables each potential participant to decide whether a project is worth pursuing, and it serves as a basis for negotiating commitments to execute the project and distribute its benefits among the participants.

Assessing Additionality and the Baseline

The first steps in the scoping exercise are to assess a project's additionality and its baseline. Additionality ensures that the project's approach is a true departure from business as usual so that the offsets claimed for the project are in fact real. The baseline is composed of the emissions and changes in carbon stocks that would have occurred in the absence of the project. Recall that developers subtract the baseline from the changes in emissions and carbon stocks they actually achieve on project lands to determine the project's net GHG benefits. The principles used to assess additionality and estimate a baseline during scoping are the same as those used to establish additionality and quantify an actual baseline while implementing the project (see Chapter 5). We recommend a method known as proportional additionality, which considers additionality and the baseline simultaneously, so that the offsets credited to a project are proportional to the degree that the project is additional.

Developers need to bear in mind that baseline emissions and carbon stocks that actually occur during the project may differ substantially from those found at the time of scoping. For example, suppose an animal operation is now using an open lagoon to store manure and earlier monitoring or independent studies have shown that lagoons of that size and type emit 50 tons of methane a month. A project developer wants to install an anaerobic digester, which would capture most of the methane and burn it in a flare, so the operation would emit only 2 tons of methane per month. The developer would initially assign a value of 50 tons per month to the baseline, and he or she would estimate a total of 48 tons per month in net GHG benefit from the project. However, suppose that shortly after the project begins, new regulations require all such operations to get rid of their open lagoons to protect nearby water supplies. Suppose the common approach to meeting the regulations leads to emissions of 40 tons of methane per month. Once the new regulation goes into effect, the project developer would have to adjust the baseline downward to 40 tons per month, and the net GHG benefit would fall to 38 tons per month.

Forecasting Changes in Carbon Sinks and Greenhouse Gas Emissions

Developers can usually estimate a project's net GHG emissions and changes in carbon stocks using either modeling or comparison sites, where practices that the project intends to employ are already in use. These techniques are usually different from those used to measure actual outcomes from a project. The former are used to project the amount of offsets based on anticipated conditions and results, whereas the latter reflect actual measurements and calculations made during and after the project.

Forecasting Biomass Gains from Forestry Projects

Scoping a forestry project requires projecting how much extra carbon the trees will accumulate because of the project and when the accumulation is likely to occur. In principle, regional publications documenting the average annual rate of forestry growth could provide a quick and cheap way to estimate carbon gains from afforestation projects, which grow new forests on land, or from projects that extend rotation lengths on land that is already forested and regularly logged. To be universally useful, such publications would need to provide the amount of extra carbon stored in trees each year as a function of species; the age of the stand or the size of the trees; and other variables, such as site productivity, management, and history. However, the few tables that have been published do not cover all forest types, growing conditions, and management practices, so project developers must often use other methods.

One alternative is to estimate gains in stored carbon by consulting "benchmark" studies that forecast carbon uptake for sites with similar species, history, productivity, and management. However, developers must consider how project lands differ from comparison lands and what those differences imply for project outcomes. A site where an old forest was recently clearcut, for instance, will probably have large amounts of woody debris and logging slash, which will decompose over time and add carbon to the soil. Thus a project that entails seeding and growing a forest on a site that has not been recently clearcut will likely record a lower net increase in carbon. Developers can make their scoping estimates more robust by comparing their site to other sites with both greater and lesser productivity.

Some kinds of carbon stocks will not change significantly during a forestry project. For example, the carbon gains in understory plants on sites that grow forests from bare ground will never be large relative to the carbon gains in the trees. Developers may find it cost-effective to use a general estimate of sequestration in the small stock rather than making a more precise estimate or to ignore the potential carbon gain in the small stock completely.

Developers may also use models to estimate the amount of carbon a tree-growing project will sequester. These models predict how trees in a single stand or multiple stands will grow, and they use equations to convert the increase in biomass to amounts of carbon. The simplest models predict how a single stand will develop. These models are simple to run, but users must still select the inputs to drive them and assess the reasonableness of their predictions. Developers may choose to hire a consultant to do so, or they may ask the project's eventual quantifier to perform the model calculations.

Multistand models can help account for some chance occurrences, such as wildfire, if a project entails actively managing a forest with many stands. Running the model thousands of times using probabilistic choices for whether a fire occurs or spreads provides a range of potential outcomes. However, such landscape simulations usually require a great deal of knowledge, and their outputs are often of little value without real data. Developers might use such models to analyze likely returns on substantial investments, rather than to make initial estimates of whether projects are feasible.

Models that predict only changes in the total volume of timber in a stand, rather than the number of trees of different sizes, are usually not suitable for predicting carbon sequestration. For example, models of tree growth often predict significantly higher changes in timber volume—and thus carbon sequestration—than observed in reality for old stands. Models that estimate forest conditions and carbon stocks over more than a century also have limited reliability, as small errors in estimates of tree growth or mortality can greatly influence the results over time. Expecting forest managers to pursue activities decades hence based on a plan written today is unrealistic in any case. Very long-term modeling can help rank alternative strategies but cannot predict actual results.

Estimating Gains in Soil Carbon

Changes in soil carbon result from a complex interaction of climate, soil texture, topography, vegetation, soil disturbance, and history of all those factors. Because of this complexity, no standard tables forecast the amount of carbon in soil stemming from changes in land management. However, developers can use a variety of alternative methods—and preferably at least two. If the methods yield similar estimates, they are more likely to be reliable, although their reliability may still be uncertain.

Developers can use user-friendly and well-tested computer models to forecast changes in soil carbon resulting from changes in land management. However, these models encompass only a limited range of techniques for managing lands. If the history of project lands and anticipated project activities are similar to those available in the models, the models can provide reasonably sound estimates of the changes in carbon stock that the project can achieve. If the lands and activities are not similar to those in the models, then the models are of limited utility.

Project developers can also consult soil scientists at long-term research sites and agricultural experiment stations. If the types of sites and land-management practices these scientists study correspond to those anticipated for the project, the results can provide a benchmark for estimating the likely increase in carbon stocks on project lands from those activities.

If planned land-management practices have been in use on a similar site for a number of years, a simple comparison of soil carbon levels between that site and the project site can prove useful. For example, if a farmer plowed an adjacent field for decades and then switched to no-till cropping, any difference in carbon between that field and the project site could indicate the gain that might accrue from a project involving no-till farming.

Estimating Changes in Methane and Nitrous Oxide Emissions

The approach to forecasting changes in methane and nitrous oxide emissions on project lands is similar to that used to quantify actual changes in emissions during the project, except that scoping entails estimating rather than observing some site conditions. The steps in this process include specifying the baseline management regime and the project regime and then comparing the stream of emissions expected to result from each (see Chapter 9).

Accounting for Emissions from Project Operations

In estimating how many offsets a project will create, developers need to take into account the inadvertent GHG emissions their land-management activities may produce during a project. For example, almost all projects produce fuel emissions, and some produce nitrous oxide emissions from fertilizer use.

Developers can start by quickly estimating fuel emissions from activities such as preparing the site, applying fertilizer, spreading manure on fields, thinning trees, and spraying from aircraft, as well as from the monitoring and verification activity itself. If total fuel use is large and varies substantially by activity, developers should calculate fuel emissions for each activity more precisely.

Estimates of fuel use for heavy equipment are based on the amount of fuel consumed per hour rather than miles per gallon. Equipment operators should estimate the number of hours required to perform each project activity (pilots should include the time needed to fly to and from the project location) and multiply the total number of hours by the amount of fuel used per hour. Developers can then find the tons of CO_2e that the amount of fuel would emit (for more on inadvertent emissions, see Appendix 2). To estimate emissions from the use of additional fertilizer, developers can use standard equations (see Appendix 2 and Chapter 9). Developers can then use global warming potentials to convert N_2O emissions to CO_2e (see Table 2.1).

Manufacturing fertilizers produces substantial GHG emissions because the process is energy intensive. However, because these emissions occur upstream from the project, developers do not have to include them when estimating the project's emissions. Under a mandatory emissions cap, those upstream emissions would count against allowances credited to the fertilizer manufacturer. In an uncapped situation, the emissions could be seen as a kind of leakage (see below), and developers might need to subtract those emissions from the project's net benefits.

Predicting a Project's Leakage

If a project reduces the amount of some good or reduces the amount of land in some use at a particular location, it may increase emissions elsewhere. Those emissions are referred to as leakage, and must be subtracted from the project's net GHG benefit. For example, if landowners plan to stop harvesting timber from a forest, they should estimate how much harvesting may occur elsewhere as a result, along with the greenhouse emissions from such harvesting. Similarly, if a project removes land from cropping, developers should estimate the amount of land elsewhere that is likely to be drawn into cropping in response to the project. The methods used to estimate leakage in the scoping phase, involving economic tools and models, are similar to those used during the quantification phase (see Chapter 10).

Intensifying production on project lands to sequester carbon can increase GHG emissions from some sources but should not cause leakage. For example, intensifying forest management generally means using more fuel, thus producing more operational emissions. However, such an approach should not increase emissions from other forests, and therefore it should not produce leakage. On the other hand, intensified management can increase upstream emissions from manufacturing inputs (such as fertilizer) or downstream emissions from the use of forest products. In a capped situation, developers would not count those emissions as leakage because they would not own or control them. However, in an uncapped situation, a developer could treat them as leakage.

Assessing Risk

Scoping a project should include assessing risk—that is, the likelihood that some portion of the anticipated offsets and financial returns will not materialize. Sources of risk include

–*Poor counterparty performance:* A project developer may not have the capacity to execute project activities or maintain project lands, especially if the project is planned to last decades.

–*Production shortfalls:* Even if participants do exactly what they plan, a project may generate fewer tons of offsets than expected, perhaps because of unanticipated weather, crop failures or disease, fire, or flooding.

–*Price changes:* Costs may be higher—or revenues lower—than anticipated. For projects with components priced in different currencies, the value of those currencies may fluctuate.

–*Errors in baselines and leakage:* Estimates of baselines and leakage may be too low, so a smaller proportion of GHG benefits may count as offsets.

–*Large uncertainties in calculated offsets:* Despite the best efforts, the methods used in the project may yield relatively large statistical uncertainty in the calculated offsets.

–*Faulty measurement and sampling:* Poor design or execution of project measurements may mean that quantifiers fail to detect some carbon sequestration or cuts in GHG emissions.

–*Regulatory uncertainty:* Regulators or verifiers could conceivably disqualify offsets. However, regulatory systems and good certification systems should have mechanisms for reducing this risk, requiring that developers receive preapproval for their project design and their methods for quantifying their results.

It is hard to overemphasize the fact that events may not transpire as planned. For example, even if models of tree growth are accurate, the amount of carbon a forestry project sequesters can differ significantly from the estimated amount if some areas prove unsuitable for planting trees or if stands take longer than usual to become established. Developers should therefore estimate high and low ranges of possible outcomes and base their risk assessment on the expected value, or the average value of all possible outcomes.[1] Of course, good information for projecting the likelihood of different outcomes is often not available, so such analyses may be subjective.

Weighing the Bottom Line in Project Offsets

Once developers complete the steps above, they can use Equations 4.1 and 4.2 to estimate the offsets they expect a project to produce:

Expected Offset = Net GHG Benefit × (1 − L) Equation 4.1

where *Net GHG Benefit* is the GHG emission reduction or increase in carbon stocks estimated from project activities, less the baseline (and any proportional additionality), and L is leakage (expressed as a fraction of the *Net GHG Benefit*). Developers can further refine the expected amount of offsets by factoring in risk aversion:

Expected Offset (corrected for risk) =
Expected Offset × (1 − Risk)
Equation 4.2

where *Risk* is a weighted factor that accounts for events that might cause a project to fail to deliver some or all of the planned offsets, as well as for the anticipated level of uncertainty in the generated offsets. This factor can be tailored to the specific inclinations of individual developers. Those who are risk averse would assign greater weight to possible negative outcomes, while risk-seeking speculators might assign lesser weight to those outcomes.

Evaluating a Project's Financial Costs and Returns

After estimating the amount of offsets they expect a project to produce, developers need to estimate their cost per ton, as well as the net income the project will provide. Even if developers implement a project because they believe in it, they need to estimate these financial costs and benefits to ensure that enough funds are available to complete it and that they will reap as many offsets as possible for the money they spend.

To estimate the costs of the land-management activities they will employ, project developers can rely on commercial software packages and other published materials. These estimates need to take into account a variety of indirect costs for maintenance and transactions. If such costs are greater than the value anticipated from the sale of the offsets, the project is probably not worth doing. Maintenance costs can include the costs of patrolling a project, managing animals, controlling fire, and taking other protective actions, as well as property taxes. Transaction costs include the costs of organizing a project, negotiating contracts, quantifying and verifying results, obtaining official recognition of offsets, and transferring rights to offsets. Transaction costs can be greater than the value of the offsets that small projects generate. However, such costs will tend to shrink dramatically as more offset projects occur, and that is why developers often find it useful to aggregate offsets from multiple projects into a single portfolio for marketing.

Developers determine the cost of each ton of offsets by dividing the project's total costs by the expected amount of offsets:

$$\text{Cost Per Ton} = \frac{\text{Expected Cost}}{\text{Expected Offsets}} \qquad \text{Equation 4.3}$$

If the cost per ton obtained from Equation 4.3 is lower than the anticipated market value of the offsets, then the project has the potential to produce a net financial gain. Determining whether the project is able to produce financial gain requires analyzing its levelized cost.

Finding a Project's Levelized Cost

As noted, different projects produce costs and benefits at different times, making comparing projects difficult. Consider the following simple example.

A buyer needs 10 tons of offsets and can purchase them today from a project that requires an investment of $100 today. The buyer would be paying $10 per ton in today's money. Now suppose that the same buyer has the option of investing in another project; in this project, the buyer can receive 12 tons and pay only $90. However, the buyer would not receive the offsets until the tenth year of the project and must pay the costs in the third year.

Which project entails the lowest cost per ton? Levelizing, a technique for putting the costs and benefits of different alternatives into the same terms, can provide the answer. If the levelized cost of one project is much higher than that of another project, project developers and offset buyers can obtain more mitigation for the money by choosing the latter.

Developers and buyers can use standard methods of financial analysis to quantify preferences for deferring costs until later and obtaining benefits sooner. That approach, known as discounting, is like calculating interest, except that analysts move backward in time from the future to the present. That is, developers and buyers use discounting to find the value today, or the present value, of a payment that they must make or the income they might receive in the future.

The first step is choosing a discount rate; different project participants will choose different rates. Low rates might match interest rates for very low-risk bonds, such as U.S. Treasury bills, whereas higher rates would reflect inflation or higher risk. Because the discount rate can have a large effect on the outcome, participants would do well to use low, high, and mid-range discount rates and then compare the results.

After choosing a discount rate, developers and buyers can find the levelized cost of a ton of offsets by dividing a project's discounted costs by its discounted benefits. The result provides a basis by which they can compare different projects. For example, the levelized cost of offsets in the second scenario above—assuming an annualized discount rate of 6 percent—would be $11.68 a ton, compared with just $10 a ton in the first scenario. Thus the latter option would be more attractive (see Appendix 4 for more on how to obtain these results). These calculations are easy to perform using a spreadsheet program or a pocket calculator with financial functions. Because of compounding, even a modest discount rate over a decade or more makes the present value of a future cost much less than today's cost.

Deciding Whether to Proceed

These analytical processes can provide a great deal of useful information on the amount, timing, costs, and net present value of offsets a project is likely to generate. Such analyses may show that a project's expected return is modest relative to the risk or that the return is lower than investors require.

Prudent developers will also consider other factors

in scoping a potential project. Landowners, for example, are often concerned about long-term restrictions on the use of their land, as well as the net return from the current use of their land. For example, a contract for producing offsets could limit the extent of logging, which could defer or permanently limit revenue from wood products. An agricultural project could limit tillage, reducing options for dealing with weeds, soil compaction, and high levels of crop residue, which could threaten revenue from crops. If creating and maintaining offsets reduces landowners' income or flexibility, the payment they receive must be enough to offset that lost income and flexibility. Even if the gross revenue from creating offsets is high, the project may yield lower net revenue than an alternative use if the project's costs are high and the project prevents another revenue-producing use. If a contract to produce offsets obligates a landowner to deliver a certain number of tons of offsets rather than to pursue specific land-management practices, the landowner must also factor in the cost of obtaining replacement offsets should the project fall short.

Offset projects may also create nonfinancial benefits, such as promoting biodiversity and generating goodwill. Potential participants might therefore weigh a project's environmental and social benefits against its difficulty and expense. For example, farmers may expect to reap only a small payment for storing carbon by switching from plowing to no-till farming, but they may contract to do so because they judge no-till farming to be more sustainable.

Chapter 5

Step 2: Determining Additionality and Baselines

To create offsets, a landowner or project developer must show that a project has actually produced greenhouse benefits beyond those that would have occurred under business as usual (practices that landowners would have pursued if the project had not occurred). Making such a determination is of course challenging, as it is difficult to know what might have happened if history had played out differently. Nevertheless, reasonably objective methods can be used to make that determination.

Such an effort typically involves two tasks. The first is to establish that a project is additional—that the land-management practices it pursues represent a true departure from business as usual. The second is to establish the baseline—the GHG emissions and changes in carbon stocks that would have occurred on project lands if the practices had not been adopted. To determine a project's net benefits, quantifiers later subtract project emissions from baseline emissions (i.e. subtract baseline carbon stocks from carbon stocks recorded during the project).

The method recommended here is based on *proportional* additionality. This approach treats additionality and the baseline simultaneously so that the offsets awarded to a project reflect the proportion of the project that is additional. (For details on these steps beyond those in this chapter, see Appendices 5 and 6.)

In contrast, many regulatory systems use *categorical* tests to establish additionality, including the Kyoto Protocol's Clean Development Mechanism (CDM),

which allows developed countries to buy GHG offsets from developing countries. If the project does not meet the additionality test, it is disallowed as an offset project. If the project does meet the test, analysts use a second step to establish a baseline. (See Appendix 5 for the processes used to establish categorical additionality.) Obviously in the case where existing regulatory and voluntary systems for reporting or limiting GHG emissions exist, projects need to follow the requirements of those systems.

Setting Baselines Using Proportional Additionality

The fundamental concept underlying the recommended method for establishing additionality and setting a baseline is that, in the absence of the project, project lands would have been managed like comparable lands in the region. Thus outcomes on other lands provide the benchmark for measuring the GHG benefits, or offsets, produced by the project. This approach depends on identifying appropriate comparison lands and quantifying GHG fluxes and carbon stocks on those lands.

The first step is to identify lands that are comparable to project lands and representative of land-management practices in the region. The second step is to see what happens on these comparison lands as the project proceeds. The difference between the emissions and sinks on comparison lands and those on project lands represents the GHG benefits that can count as offsets. For example, if comparison lands gain 1 ton of carbon per acre and project lands gain 3 tons, 2 tons per acre may count as offsets. This method for calculating the baseline uses the proportion of the GHG benefits achieved by the project that exceeds any benefit achieved on comparison lands.

Identifying Comparison Lands

Project developers typically select comparison lands after they decide to proceed with a project but before its land-management activities begin. Comparison lands ideally resemble project lands in their physical characteristics, including weather, soil, and topography. Land-management practices on comparison lands at the outset of the project should also roughly mimic those used in the region so these lands can serve as a valid measure of the degree to which the project exceeds business as usual. Changes in land-management practices on the comparison lands throughout the project period then represent the region as a whole. The baseline GHG emissions or changes in carbon stock measured on the comparison lands during the project provide a continuous update of the project's additionality.

An alternate approach is to choose comparison lands that are subject to the same uses and practices as the project lands before the project starts. Project developers assess the fractional additionality of the project at the outset and discount the GHG benefits they later achieve by this amount. For example, suppose a project plans to pursue no-till farming of small grains, and 40 percent of the farmers of small grains in the region already use no-till farming. In this approach, project developers would discount the amount of GHG benefits the project achieves by 60 percent. If the comparison lands shift to no-till farming during the project, the baseline would reflect those changes.

Field surveys of land cover on potential comparison sites can sometimes help identify the management activities on those lands. For example, the proportion of land covered by residue from an earlier crop after a new crop is planted usually indicates the degree of tillage. However, project developers should try to choose comparison lands based not just on land cover but also on

known management practices.

Comparison lands should usually be near project lands to ensure that they have similar weather, topography, and ecology. Forestry regulations vary widely from state to state, so project developers would do well to choose comparison lands within the same state. If ecological conditions in an area vary greatly, developers should also choose comparison lands based on elevation, precipitation, soil type, and productivity. Many states tax resource lands according to their potential productivity, and official maps often include information on those attributes. However, project developers need to keep in mind that comparison lands with similar productive capacities and subject to the same regulations may face very different market opportunities (including proximity to transportation, ownership structure, and economic risk), and they should choose comparison lands accordingly.

To avoid the cost of performing annual surveys to determine changes in the uses of comparison lands during a project, developers should seek lands subject to an existing survey. Several such surveys are available for U.S. lands:

– The National Resources Inventory of the U.S. Department of Agriculture (USDA) includes information on land use and land cover, crop history, and conservation practices for 800,000 sites on federal land. Each site represents a larger land area. Although information is not updated every year, the agency is moving toward an annual inventory. (See http://www.nrcs.usda.gov/technical/NRI/.)

– USDA census data, collected on a five-year interval (mostly since 1974), includes county-level information on farm size, livestock numbers, and crop acreage by irrigation status. (See http://www.nass.usda.gov/census/.)

– The USDA's Forest Inventory and Analysis reports on forest cover, growth, mortality, tree removals, and general health for all forestland in the United States. The information, updated every 5 to 10 years, is reported mostly at the state and county level to protect the confidentiality of landowners. (See http://fia.fs.fed.us/.)

– Purdue University's Conservation Technology Information Center provides annual information at the county level on tillage practices by crop. Analysts can use the information to determine the overall use of conservation tillage, but not the probability that land will change from one use to another over time. That is because the data do not show whether specific land parcels remain subject to a given management practice from year to year. (See http://ctic.purdue.edu/CTIC/CRM.html.)

– The National Land Cover Mapping of the U.S. Geological Survey shows 21 classes of land cover at the 30-square-meter scale, based on satellite imagery from the LandSat Thematic Mapper in 1992 and 1999. (See http://edc.usgs.gov/geodata/.)

If information on specific uses of comparison lands is not available, project developers may have to rely on general categories of land use (such as annual crops versus pasture, rather than variations in tillage practices used to cultivate annual crops). However, that approach may increase the risk that comparison lands will differ from project lands in some unmeasured way.

Determining Baseline Emissions and Sequestration

A robust approach to establishing baseline GHG emissions and carbon stocks on comparison lands is to employ the same method used to quantify emissions and stocks on project lands—and to do it over the same time period. For example, quantifiers for a project that aims to sequester carbon in soil could directly sample carbon stocks on comparison lands, as that is the approach they will use to measure carbon gains on project lands. However, gaining access to comparison lands to make such measurements is difficult. To circumvent that problem, developers can set aside a small fraction of project lands or facilities where they do not implement project activities. They can then use these particular lands to help characterize baseline GHG emissions and sequestration. Still, even this will not suffice if land-management practices on comparison lands change during the project period or if the region has an array of practices (see the next section).

Cost can be an even bigger impediment than access to comparison lands in establishing baseline GHG emissions and carbon stocks because these lands can include an array of land-management practices. Determining the baselines may then be more costly because measuring changes in emissions and stocks may require taking more field samples and even using stratified sampling (see Appendix 21).

To avoid the challenges of taking field samples on comparison lands, some carbon-trading systems may allow quantifiers to use very limited information on conditions and practices on those lands, along with other information or models, to estimate a baseline. For example, Birdsey and Lewis (2003) employ models to calculate trends in stocks of forest carbon in each state, which quantifiers could use to establish baselines for forest projects. The Intergovernmental Panel on Climate Change provides default emissions rates as a function of land-management practices for methane projects. The USDA's Natural Resources Conservation Service offers the COMET-VR tool that estimates soil carbon stock change for selected cropping practices in the U.S. that could be used to establish project baselines for soil carbon projects. In another approach, the USDA or the U.S. Forest Service could periodically (perhaps every 5 or 10 years) publish default baselines for various types of offset projects as a function of their location and topography.

Although such information and models provide a relatively inexpensive and convenient way to estimate baselines, they will generally be less accurate than direct sampling of comparison lands. The carbon-trading system or offset buyer will determine whether less expensive but less accurate methods are acceptable. Even in trading systems that do not require sampling to establish a baseline, savvy buyers may insist that analysts use robust methods. The resulting offsets should bring a higher price than those with baselines based on less accurate methods, and project developers may decide that the higher price justifies the extra costs.

Quantifiers must express baselines in the same units they use to express changes in GHG emissions or carbon stocks on project lands. For example, if they express project emissions and stocks as total tons of CO_2e, and they first calculate the baseline as tons of CO_2e per acre or per unit, they must multiply the baseline by the number of acres or units encompassed by the project.

If the project involves emissions cuts, then quantifiers subtract baseline emissions from project emissions to obtain the net GHG benefit. If the project involves sequestration, then quantifiers subtract baseline carbon stocks from project carbon stocks. Quantifiers then adjust this net greenhouse benefit for leakage (see Chapter 10). Once ownership is established and the benefit is verified, it may count as offsets and be marketed.

Timing is another issue. A project is typically divided into accounting periods during which offsets are calculated and marketed. For example, a project that lasts 10 years could establish 10 one-year accounting periods. Quantifiers must determine the baseline for each accounting period.

Accounting for Changing Baselines over Time

In practice, GHG emissions and sequestration rates for a given tract of land can change as a result of changing environmental conditions and management practices. For example, low crop prices can prompt farmers to abandon agriculture on marginal comparison lands, and the cessation of plowing and reestablishment of woody vegetation can boost soil carbon stocks. A baseline can change even when the mix of activities on comparison lands does not. For example, intensified logging on comparison lands can lower stocks of biomass carbon, or better livestock management can help restore riparian conditions, which can increase woody biomass and thus soil carbon. Project developers and quantifiers must therefore analyze both changes in use and site dynamics, including climate variability, on comparison lands during the project to construct reliable baselines.

Figure 5.1 shows how such changes can influence the amount of offsets an emissions abatement project creates. The project is divided into 12 accounting periods for which quantifiers calculate offsets. During the first five periods, comparison lands emit 35 tons of GHGs. However, by the last accounting period, emissions drop to just over 25 tons. Thus baseline emissions vary from 35 to about 26 tons over the course of the project.

Like emissions from comparison lands, emissions from project lands can also change for reasons beyond the control of project developers, and developers must similarly account for these changes. In Figure 5.1, emissions from project lands initially fall from 35 tons to about 22 tons during accounting periods 7 and 8, and then they rise to about 26 tons by the end of the project. The net GHG benefit for any given accounting period (before accounting for leakage) is the difference between baseline and project emissions. In this example, net mitigation on a per-acre basis is zero in the first accounting period, then peaks at just over 10 tons during accounting periods 5 and 6, and then declines to zero during the last accounting period. Quantifiers would find the net amount of offsets by multiplying the per-acre GHG benefit by the number of acres in the project.

Accounting for Variable Land-Management Practices

A further complication in determining baselines arises from the fact that comparison lands are often subject to several land-management practices. In such cases, quantifiers determine the baseline by calculating the "weighted average" of the changes in emissions or carbon stocks from each land-management practice. The equation for doing so is

$$B_A = \sum_{i=1}^{j} \frac{e_i \times \text{area}_i}{\text{area}_{\text{total}}} = \sum_{i=1}^{j} e_i \times fa_i \qquad \text{Equation 5.1}$$

where B_A is the average baseline emissions or sink per unit area, Σ is the sum of all uses i applied to comparison lands (from use 1 to use j), e_i is the emissions or sink per unit area for each use i, area_i is the area in use i, $\text{area}_{\text{total}}$ is the total comparison area, and fa_i is the fraction of the total comparison area in use i. That is,

$$fa_i = \frac{\text{area}_i}{\text{area}_{\text{total}}} \qquad \text{Equation 5.2}$$

For example, consider a project involving no-till farming that has chosen 100 acres of comparison lands. During a given accounting period, 20 acres (or 20 percent) of comparison lands are also in no-till cropping, while the remaining 80 acres (or 80 percent) use standard till-

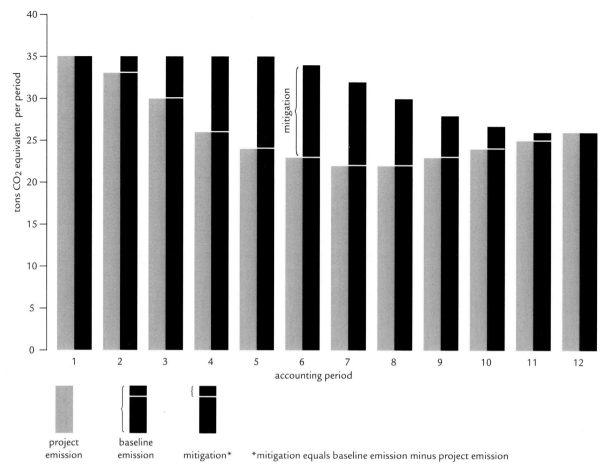

Figure 5.1 How changes in baseline and project emissions affect the tons of CO_2e that a project creates. The mitigation or net greenhouse benefit from a project is the difference between the baseline and project emissions. Because both types of emissions can change over time, both must be tracked over each accounting period. In this figure we show how emissions might change for a project that is divided into 12 accounting periods.

ing. No-till cropping boosts carbon stocks by 0.5 tons per acre during the accounting period, while tilling sequesters 0.1 tons per acre. In this case, the baseline increase in carbon stocks per acre is

$$B_A = 0.2 \times 0.5 \frac{\text{tons}}{\text{acre}} + 0.8 \times 0.1 \frac{\text{tons}}{\text{acre}} = 0.18 \frac{\text{tons}}{\text{acre}}$$

If project lands are all subject to no-till farming and sequester 0.5 tons of carbon per acre, the net GHG benefit during each accounting period equals 0.32 tons per acre (0.5 tons per acre – 0.18 tons per acre).

When comparison lands are subject to several different management practices, these calculations become more arduous, and a spreadsheet can make calculating

the weighted average much easier. Table 5.1 provides a template for such a spreadsheet applied to our simple no-till/till example.

Another complication arises from the possibility that uses and management practices on comparison lands can change during the project period. Because developers choose comparison lands to reflect what happens under business-as-usual circumstances, establishing a baseline requires them to track all changes in the use of comparison lands and to analyze the impact on GHG emissions or carbon stocks. In principle, they can do this by documenting uses and management practices on comparison lands for each accounting period, as well as changes in emissions or carbon stocks on these

Table 5.1 Template for a Spreadsheet to Calculate a Weighted-Average Baseline for Multiple Uses in the No-Till/Till Example

Land-use type	fa_i Fractional cover on comparison lands	C sequestration for land-use type over period (tons/acre)	Fractional contribution to C sequestration on comparison lands (tons/acre)
No-till	0.2	0.5	0.1
Till	0.8	0.1	0.08
B_A = Total Baseline = Total of fractional contributions =			0.18 tons per acre

lands (using Equation 5.1 to calculate B_A). However, developers are sometimes unable to track management practices on comparison lands during each accounting period—perhaps they can determine them only at the beginning and end of the project. In that case, they must estimate the impact of changes in activities on comparison lands for each accounting period using the periodic transition rates in Appendix 6.

Establishing a project baseline poses a number of challenges for developers and quantifiers, but doing so is essential to creating real and accurate offsets. In most cases, regulatory agencies and offset brokers and buyers will insist on a transparent process for developing a baseline so they can independently verify its rigor. Project developers and quantifiers should therefore give careful thought to how they will address additionality and the baseline during the scoping process, and then they should scrupulously follow their chosen method throughout the project. The most robust approach is to choose comparison lands with properties like those of project lands but that are subject to management practices at the start of the project that represent the region. The baseline is then the rates of GHG emissions and carbon sequestration observed on these comparison lands.

Chapter 6

Step 3: Quantifying the Carbon Sequestered in Forests

Forests represent significant reservoirs of carbon captured from the atmosphere through photosynthesis. If released from forests, this carbon would largely convert back into atmospheric CO_2. Reforestation—the process of shifting previously forested lands that had been converted to other uses to stands of growing trees through natural regeneration or planting[1]—sequesters carbon from the atmosphere and thus produces GHG benefits.

The amount of carbon stored in forests depends on their type, as well as the climatic conditions and management practices to which they are subjected. However, patterns of sequestration are similar among different types of forests (see Figure 6.1). Shortly after a clear cut or fire, when new trees are relatively young and small, sequestration rates are low. After trees grow to the point where they fully occupy the canopy, the rate of sequestration rises and continues at a high rate for several years. In many forests, this period of rapid accumulation of carbon persists for several decades. As the trees mature, annual growth and sequestration slow, but the cumulative amount of stored carbon is substantial. In very old forests, the amount of carbon in the stand may continue to increase slowly or may decline. In very old forests, tree death can cause large trees to become widely spaced, reducing the total carbon stock of the forest. The carbon stock in mineral soil and the forest floor can continue to increase as a result of annual litter inputs and the decomposition of woody debris. Overall carbon stocks can decline, however, if

succession produces a shift to species in which individual trees do not grow as large.

Because of these complex changes in carbon accumulation, a well-designed system for sampling forest biomass is critical to an offset project.[2] Developers must be able to accurately measure carbon sequestration without incurring prohibitively high costs. This is especially important because forest projects usually last for decades.

Sampling designs for forest projects must therefore be:

- Accurate and repeatable over long periods of time.
- Adaptable to unforeseen circumstances, such as wildfires, forest management changes, and the addition or removal of lands from a project.
- As simple as possible to allow outsiders to audit results.

This chapter describes an approach for quantifying sequestration that is designed to reduce variability, control costs, and detect much of the sequestration a project achieves. This approach is based on extensive experience in measuring changes in forest carbon and entails the following steps:

- Designing a forest sampling system that is robust with respect to the different locations of carbon accumulation.

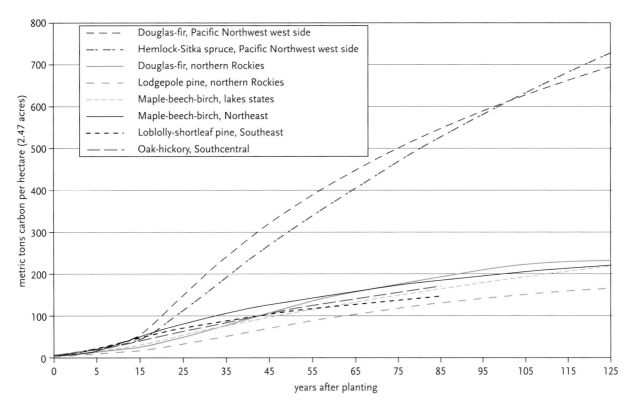

Figure 6.1 Growth in total above- and below-ground forest carbon stock, after planting land previously in non-forest cover. This includes carbon associated with live and dead woody material but excludes carbon in the mineral soil. The total amount of carbon stored in forests depends on their type, as well as the climatic conditions and management practices to which they are subjected. However, the basic patterns of sequestration tend to be similar.
Note: Calculated from amounts reported in U.S. Department of Energy 2006. 1 hectare = 2.47 acres.

– Conducting initial field measurements of the different sites of carbon stocks in a forest.
– Selecting allometric equations for converting field measurements into carbon mass, or developing new ones.
– Taking subsequent field measurements to determine changes in carbon stocks over time.

Crucial aspects of this approach include performing unbiased sampling, choosing an adequate number of sampling sites, and deciding whether and how to stratify sampling across a site. (See Appendix 2 for more on sampling, and see Appendices 7–15 for more details on the steps described in this chapter.)

Quantifiers must perform the steps listed correctly when the project is established, the first time, as they cannot go back in time and redo them. They should re-peat quantitative field measurements every five or 10 years, relying on annual qualitative or quantitative observations in intervening years to determine whether a project is proceeding according to plan and to take remedial action, if needed. As with other projects, developers should aim to detect net carbon sequestration with an uncertainty of 10 percent at a 90 percent confidence level, as the potential benefits of greater accuracy are generally not worth the added cost (see Appendix 3).[3]

To ensure that its system for quantifying carbon is accurate but not overly costly, a forest project should encompass at least several hundred hectares and generate at least 100,000 tons of CO_2 equivalent in offsets. Project developers with smaller areas, or who seek to generate fewer offsets, should consider combining their lands with other parcels.

Dividing a Forest into Carbon Pools and Using Subplots

A forest project's plan for sampling carbon stocks in the field must evaluate all types of biomass,[4] including live trees, shrubs and seedlings, standing dead trees, downed woody debris, the forest floor, and possibly mineral soil. Quantifiers will track these *carbon pools* separately throughout the project. Remotely sensed imagery can provide a helpful guide in locating the various types of pools present on a project's lands (see sidebar). If quantifiers conclude, based on existing scientific knowledge, that a particular pool will not lose or gain a significant amount of carbon, they may remove it from the sampling plan, but comprehensive field measure-

ments will be far more persuasive to independent verifiers and potential buyers. Quantifiers should certainly measure pools that are likely to lose carbon, to avoid accusations that their analysis is biased. (See Appendix 7 for more on carbon pools.) Attention should also be paid to deciding whether mineral soil carbon stocks should be measured. Scientific knowledge should be used to predict whether project activities have a significant chance of causing a decrease in mineral soil carbon. If so, mineral soil carbon should be measured (see Chapter 7 for methods for measuring change in mineral soil carbon stocks),

To measure biomass carbon, field crews first create an adequate number of unbiased located sampling sites, or plots.[5] Then, within each plot, field crews lo-

Using Imagery to Design a Carbon-Sampling Program

In all but the simplest projects that aim to sequester carbon in forests or soil, detailed remotely sensed imagery provides key information for designing and executing an efficient system for measuring changes in carbon stocks. Images can help delineate the project area, define the extent of project activities, and group similar areas together, thereby increasing the precision of measurements of carbon sequestration.

Several types of remotely sensed images are available:

– *Orthophotos* (in either hard copy or digital format). Orthophotos provide the best tradeoff between high resolution, timeliness, and limited cost. Orthophotos have a uniform scale because they correct for parallax, enabling quantifiers to calculate the size of areas subject to specific activities. Orthophotos also typically show latitude and longitude coordinates or state plane coordinates. Such *geo-referencing* allows quantifiers to calculate the coordinates of specific locations and then use a handheld GPS receiver to travel to those locations on the ground, or vice versa.
– *Aerial photographs.* Standard aerial photographs taken on 9-inch-square negatives have high resolution and can help reveal which areas

are alike and which are different. Such photos are available for most of the United States, starting in the late 1930s.
– *High-resolution satellite images.* Satellite imagery comes in very different levels of resolution, many of which are too coarse for use in quantifying carbon sequestration, and it is often very costly. Analyzing such images require more skill and software than do aerial photos. The ability to automate analysis makes satellite images a very useful tool for use in larger projects. However, because they are taken on a weekly to monthly basis, satellite images are much more likely to capture a project closer to its start date than aerial photos. Such images are useful in tracking land-use changes (such as distinguishing annual cropping from pasture, forest, and development) or in recording wind or fire disturbance.

Maps are an alternative source of spatial data that can help users document the general location and, sometimes, the sizes of land parcels enrolled in a carbon sequestration project. Seldom can administrative/ ownership boundaries be inferred from maps, unless they were created for this purpose. Maps must be detailed enough to show land attributes such as elevations, streams, roads, and administrative boundaries; scales coarser than 1:25,000 are of limited use.

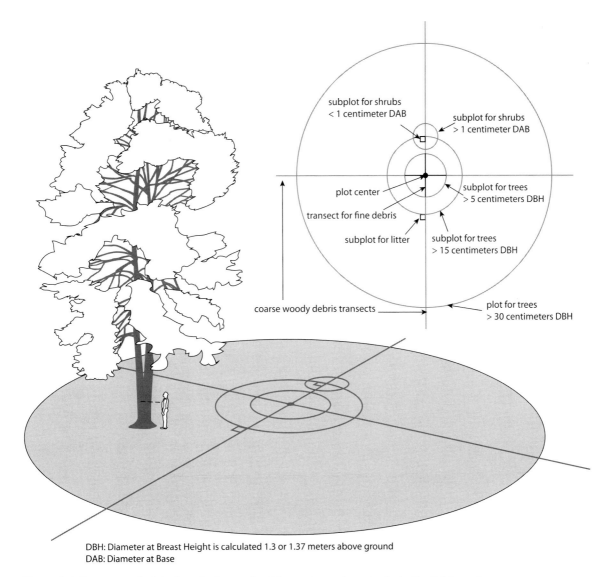

subplot for shrubs
< 1 centimeter DAB

subplot for shrubs
> 1 centimeter DAB

plot center

transect for fine debris

subplot for trees
> 5 centimeters DBH

subplot for litter

subplot for trees
> 15 centimeters DBH

coarse woody debris transects

plot for trees
> 30 centimeters DBH

DBH: Diameter at Breast Height is calculated 1.3 or 1.37 meters above ground
DAB: Diameter at Base

Figure 6.2 Recommended design of a forest plot. Measuring carbon stocks in the field requires evaluating all types of biomass, including live trees, shrubs and seedlings, standing dead trees, downed woody debris, the forest floor, and possibly mineral soil. Here we illustrate the design of a forest plot to do this. If the project encompasses areas where litter or soil O horizon exists or can accumulate, this horizon should also be measured. A project may also choose to measure carbon in mineral soil. See the chapter on measuring soil carbon for methods of quantifying changes in amounts of carbon in these pools.
Note: DBH = diameter at breast height, typically defined as the point 1.3 or 1.37 m above the ground.

cate several circular subplots—one for each type of biomass, or carbon pool—around a single point known as the plot center (see Figure 6.2).[6] Crews then make measurements specified in the field protocol of the monitoring plan. They then measure woody debris along perpendicular lines that extend in each cardinal direction from the plot center because such debris can be affected by trampling. (If the project expects to store a large proportion of carbon in woody debris, quantifiers may want to extend the length of those transects.) Calculating carbon stocks accurately requires determining whether plants (or other materials being measured) near plot boundaries are in or out of the plot, which necessitates establishing plot boundaries precisely and accurately.

Quantifiers should size plots to measure larger trees

precisely (larger trees generally being greater than 15 cm dbh in most forests or greater than 10 cm in diameter at breast height (dbh) in forests with smaller trees), as they will sequester the most carbon in most projects. Experience has shown that plots with as few as four trees can provide an accurate and precise sample of the amount of carbon a project is sequestering, even if tree size varies. However, large trees are more widely spaced than small trees, so quantifiers will need larger plots to precisely sample them. If a project is designed to grow a mature forest, and such a forest includes at least 40 large trees per hectare, a 0.1-hectare plot would probably be efficient. Trees in a natural forest are often located in clumps and at a density of 40 trees/ha. This clumpiness means that a 0.1-hectare plot would have a probability of less than 0.5 of actually encompassing four trees. However, in most forests (not woodlands savannahs), the density of moderate and large trees is usually greater than 100 trees/ha, and even with a clumpy distribution of trees, a 0.1-hectare plot would usually encompass at least four moderate- or large-size trees. If a project encompasses more than 1,000 small trees per hectare, 0.02-hectare plots might work, as they would average five evenly spaced trees. If few trees will ever get larger than the small-size category, it may require only a little more effort to make the plots a bit bigger. Having more trees per plot would substantially increase precision. However, if the project is expected to grow larger trees, it may not be worth any additional effort to get a more precise measurement of the carbon stocks in small trees. Somewhat larger plots might be efficient for sparsely vegetated woodlands or natural, spatially heterogeneous forests.

Statisticians often maintain that many small plots provide greater statistical power than fewer large plots, given a homogeneous population of objects being measured. However, in practice, maximizing statistical power for a given cost usually means establishing more large plots. This is because the cost of traveling from one plot to the next can be substantial. Spending that effort on measuring larger plots instead of a greater number of plots can yield greater precision. For example, in Table 6.1, travel time between plot centers would have to be just less than four minutes to drop the cost of establishing and measuring 0.01-hectare plots below the cost of measuring 0.1-hectare plots. This short

travel time is not feasible for widely spaced plots or for those located on noncontiguous parcels. Note also that the time to travel between plots would have to fall to just over one minute for 0.001-hectare plots to become more cost-effective than 0.01-hectare plots.

The effort and expense of measuring each carbon pool should be commensurate with the amount of carbon it is expected to sequester over the course of the project. Relatively imprecise measurements of pools with small changes in carbon stocks will have little impact on the precision of the overall measurement.

Consider a hypothetical project that expects to sequester 100,000 tons of carbon. Suppose this forest stores carbon in large and small trees only; the large trees are expected to store 90,000 tons of CO_2e, and the small trees are expected to store 10,000 tons of CO_2e. Because quantifiers expect the large-tree pool to contain roughly 90 percent of the sequestration, they should devote roughly 90 percent of the sampling effort to that pool.

Similarly, because the forest floor usually does not gain much carbon in most forest ecosystems, quantifiers may choose inexpensive methods to measure it, even if they are not very precise. For example, crews could measure the combined thickness of duff and litter[7] at one specified point on each plot. Then the density of the litter and duff could be used to estimate the forest floor mass on each plot.[8] However, the litter and duff density should be measured for each project, taking into account that it can vary significantly from season to season. If a project may have a change in the forest floor carbon stock that is a substantial fraction of the total carbon stock change within the project area, it is strongly recommended that forest-floor mass be directly measured by weighing material collected on subplots of fixed size, not inferred from thickness.

Quantifiers may decide to stratify a carbon pool across a project area or across physical characteristics to decrease variability. If there is a known difference in the physiographic characteristics (e.g. soil drainage, soil parent material, and forest composition) it is useful to stratify the project area by these variables and calculate the carbon stocks independently for each stratum. This approach reduces the total uncertainty in the final stock estimates with no additional sampling. However, stratifying requires establishing more boundaries

Table 6.1 The Cost of Estimating Carbon Stocks on Plots of Different Sizes

Plot size	Number of plots	Time to measure one plot	Total field cost
0.1 hectare	10	1.5 hours	$700
0.01 hectare	95	0.1 hours	$1,330
0.001 hectare	288	0.02 hours	$3,110

Notes: These estimates represent typical costs in lightly roaded areas of the United States. Assuming the total bundled cost of a field technician is $40 per hour, the time to get from one plot to the next and establish or re-locate the plot center is 0.25 hours per plot, and the amounts of time to measure a plot of each size are as given in the table. The number of plots of each size is based on observed variability in an unmanaged, second growth stand of mature natural regeneration in the Pacific Northwest, with the numbers of smaller plots set to yield the same statistical confidence interval as observed for 10 plots of 0.1 ha in size. *Source*: Gordon Smith, Ecofor.

and analyzing separate sets of data. Stratum boundary choices depend on the frequency of the occurrence of trees or other objects being measured, the size of subplots, the time needed to measure and analyze each subplot, and the sequestration likely to occur within each class of biomass. It is generally efficient to divide woody debris into two or three classes based on size, and to divide standing vegetation into at least three classes.

Another approach to measuring biomass is to sample a given proportion of the project area. Experience shows that sampling 1.5 percent of the project area can provide reliable measurements of forest carbon if the plan calls for several dozen sites. This approach is best for moderately sized projects of 300 to 1,000 hectares. Quantifiers can base the percentage of the project area to sample on the size of the subplot for the carbon pool expected to record the largest change.

For a plot size of 0.1 hectares, sampling 1.5 percent of the project area would mean installing one plot for every 6.6 hectares. If large changes in carbon stock are expected to occur in a pool other than large trees, obtaining the needed precision may require more intensive sampling. Quantifiers may find it more efficient to expand the size of the subplot used to sample that pool rather than increase the total number of plots because the latter approach would require more overall effort. For projects larger than 1,000 hectares, installing one plot every 6.6 hectares would require more than 150 plots. If several strata are sampled separately, it may be feasible to measure no more than 15 plots in each stratum, if the total number of plots is sufficient to achieve the desired level of precision (see Appendix 1).

After determining the number of plots, project developers should evaluate whether the measurement system will generate precise enough data to yield enough sequestration (once uncertainty is considered) to make the project economically worthwhile. If the answer is no, they can investigate whether a different level of precision would make the project financially viable. Independent verifiers should check the project's sampling approach and financial structure to determine whether the project is likely to fulfill its commitments.

Installing the Sampling Plots

Field crews must establish permanent sampling plots so crews can return decades later to remeasure the amount of carbon on each plot. Using a GPS receiver to record the coordinates of plot centers and place permanent markers is essential. A mapping-quality GPS receiver (which should be accurate to 1 to 10 meters) should enable field technicians to find the monument that marks a plot center later, although GPS measurements will be less accurate under heavy forest canopy and in narrow valleys.

To mark the plot center on sites where significant soil disturbance is unlikely, crews can drive a piece of rebar 1 to 2 feet into the ground. Fire, tree fall, and vehicle traffic will usually not disturb the rebar if it is flush with the ground, and later crews can use a metal detector to find it. For sites where significant soil disturbance is likely, crews can bury a magnetic ball marker 0.5 meters deep to mark the plot center. If major ground disturbance is likely, crews should establish two additional monuments, using a GPS receiver to record their distance and direction from the plot center. (For specific steps in installing field plots, see Appendix 8.)

Plot centers should also be recorded in a geographic information system (with a scale of 1:12,000 or larger) to help crews reach the vicinity of plot centers later. Narrative descriptions of how to find plot centers from a landmark can be useful, although things often change, making descriptions hard to follow decades later.

Choosing Resampling Intervals

Field measurements of forest biomass inevitably entail error. For example, measurements of tree diameter by two different field technicians (even if they are skilled) are likely to vary by up to 1 cm dbh in larger trees. Quantifiers can minimize this problem by lengthening the interval between field measurements, as changes in carbon stock will vastly exceed the uncertainty attributable to measurement errors. Measuring carbon stocks every five to 10 years also averages out annual variations in sequestration and allows quantifiers to detect a greater proportion of sequestration while reducing cost.

However, more frequent measurements can help project developers remedy any shortfall in sequestration that field data indicates, for example as a result of the invasion of low-carbon-sequestering species. In addition, as the time between measurements grows, so does the cost of waiting to quantify the increase in carbon sequestrated. Thus, at some point, expanding the time between field measurements becomes counterproductive.

The optimal period between measurements will vary with their precision, the speed of change in carbon stocks, the costs of measurements, and the value of the resulting offsets. Larger projects may want to measure carbon stocks every five years for the first few decades and then less often as quantifiers gain information on how much carbon a project is likely sequester moving forward, especially if sequestration rates are declining. Projects usually schedule a field measurement shortly before they end to determine total project carbon.

Quantifiers should make annual observations, either visually or using remote sensing, between more detailed field measurements, to detect major deviations from expected conditions. For example, scanning a landscape from a high point can reveal whether it is substantially covered with healthy trees. Observations of the "leader" stems of young trees can also reveal whether they are growing vigorously. If large areas show discolored foliage or if many trees are dead or missing, quantifiers can conduct detailed measurements of biomass. If projections of how much carbon a project will sequester are conservative, an observation that 25 percent of the project area is not in a healthy, growing condition might trigger remedial action. If projections are less conservative, the threshold for remedial action may be as low as 5 to 10 percent of the project area. These annual checks may be qualitative assessments or quantitative stocking surveys, such as those performed to measure the survival of planted seedlings (see Appendix 9).

Satellite imagery can be used to measure leaf area and estimate growth. However, this kind of analysis requires multiple, fine-scale images through the growing season and a skilled analyst. The costs of data and analysis necessary to estimate growth rates from satellite imagery may be more than the cost of ground-based assessments. These costs and the capacities of the quantifier will determine whether it is most cost-effective to assess vegetative condition using satellite imagery, aerial photographs, or ground based surveys.

Some offset contracts require developers to model future tree growth partway through a project to determine whether it will achieve its goals. Quantifiers should use such a model only if it has been validated for the project's location and forest type. Validation requires running the model for locations not used to build the model and for which independent data exist. Because each model has its own idiosyncrasies, modelers should have experience running the model they will use.

Modeling usually requires collecting more information than quantifying biomass. If a model requires extra information, crews should collect it from a subset of the trees used to calculate biomass. A model may also require historic information on tree growth and management activities, which can be gleaned from land records and interviews with previous managers. If historical information is unavailable, the modeler must start from current stand conditions and be aware of how this lack of knowledge could affect model accuracy. To accurately estimate future carbon stocks, the model's input should require knowing trees by species, height, and diameter, not just the volume of growing stock.

Measuring Carbon Stocks on Subplots

Determining the amount of carbon stored in the project area entails documenting the physical characteristics of objects or materials measured on the subplots, including the heights and diameters of trees and shrubs, and the mass of organic material on the forest floor. Fieldwork may include collecting samples from subplots and measuring the weight of the material. This material is then analyzed further in a laboratory. Laboratory analysis may be limited to drying the samples and finding their weight, or it may involve determining the carbon content of each sample.

Quantifiers use biomass equations to convert the gathered information into the amount of carbon in each pool per hectare. Different biomass equations use different characteristics of the trees and pieces of woody debris on the subplots (and the carbon contents of different parts of these objects) to derive total carbon content. Quantifiers must therefore identify the specific biomass equations they will use before the project starts so field crews will know what kinds of measurements to make. The next section suggests a protocol for each of these steps. A different sampling strategy would use different protocols, and could be equally valid.

Making Field Measurements and Gathering Samples

Because of the sheer variety of objects, materials, and carbon pools that a forestry project must monitor, as well as the size of some of the objects, making field measurements is challenging. Crews should measure the carbon pools in each subplot in a standard sequence, concentrating on subplots for forest floor and fine debris first, as those are sensitive to disturbance from trampling. Adhering to a standard sequence reduces the chance that the crew will forget to measure a subplot or to measure a tree on a large subplot. A standard pattern for taking measurements also helps quantifiers check them for quality control.

In making all these measurements, each crew will adapt its division of labor to its skills and the types of biomass on a plot, although one person usually records all the data. In a two-person crew, one person can measure woody debris and litter while the other person measures trees. If two people examine trees, one person can determine which trees to measure and measure diameters while the other person measures heights, determines vigor, and records data. In a three-person crew, one person can measure litter, woody debris, and small live material while the other two people measure larger trees and snags.

Fieldwork to remeasure carbon stocks later in the project resembles initial measurements, except that crews re-locate plots instead of installing them. If a layer of decomposed organic material was present above mineral soil during the first measurement, crews should remeasure this material at different locations to avoid the disturbed areas. If a crew cannot re-locate a plot, it should make its best guess as to where the plot should be and establish a new one at that location, noting the change in field records. Quantifiers can judge whether to use the new measurements when analyzing the data.

Most projects will focus most intensely on the amount of carbon in living trees. To accurately determine tree growth, and thus changes in carbon stocks, crews should follow U.S. Forest Service procedures for measuring the diameter and height of trees over 5 cm dbh. Crews measure smaller trees, saplings, and shrubs at the base.

Because small pieces of woody material and decomposed organic material will never provide a significant source of carbon, crews can count the number of pieces of a particular size rather than measuring their exact diameter or length. Quantifiers can then calculate the biomass within each class based on the median size.

To measure the amount of biomass in litter, crews gather loose leaves, twigs, bark, seeds, and other identifiable plant parts that accumulate on the ground above mineral soil up to a threshold size, and they weigh a representative subsample from each subplot. These subsamples are then dried and weighed to find the ratio of dry weight to wet weight. Quantifiers use this information along with biomass equations to quantify the amount of carbon (see below).

If litter (i.e. soil O horizon) is present or is expected to become present during the course of the project, the O horizon should be measured separately. In locations where O horizons form, the O horizon carbon stock can become very large over time and can be lost quickly though disturbances such as fire or logging. If an O horizon will be measured, it is recommended to measure duff with the O horizon, not with litter. One method for measuring mass of organic matter above mineral soil is to obtain a round template of known area (225 cm^2 is a favored size), place it on the ground at the point to be sampled, cut around the template, lift the organic material, place it on a plastic sheet, and carefully remove any mineral soil from the sample of organic material. The entire sample may be bagged and removed for drying and weighing, or the sample may be weighed in the field and a subsample removed and weighed in the field, and the subsample taken to a laboratory for drying and re-weighing to establish the dry to wet weight ratio to be used to calculate the dry weight of the whole sample.

Forest soils may also comprise a significant carbon pool and thus should be measured (see Chapter 7). Calculation of the mass of woody debris requires information on the density of material of various degrees of decomposition (see Appendix 10).[9] When forested lands are hilly or mountainous, quantifiers must correct for these sloping land features in their area calculations or instruct crews to install sampling plots in the horizontal plane (see Appendix 11).

Analyzing Biomass Samples in the Laboratory

For most species, the concentration of carbon in whole trees is very close to 50%. As a result, it is acceptable to assume that the concentration of carbon in live tree biomass is 50% and not measure the concentration. In nontree biomass (such as leafy annual vegetation) and decomposed material, the concentration of carbon is often significantly different from 50%, and the concentration should be obtained from a published source or by laboratory measurement of samples from the project area.

To determine the concentration of carbon in the samples collected, each sample must be analyzed for its dry weight and carbon content. Quantifiers with technical expertise and access to laboratory facilities can perform this analysis themselves. However, engaging a qualified laboratory will often prove less costly. In the United States, many university labs provide analytic services for a fee, as do some commercial labs. (Projects in less industrialized countries may not have access to analytical equipment.) The cost of analysis is generally a few dollars per sample—higher if more sample preparation is needed, and lower if more samples are run. Forestry and agricultural extension professionals should be able to point quantifiers to nearby labs that can analyze the chemical content of organic materials or soil. Some commercial laboratories that focus on soil nutrient testing, and many laboratories in developing countries, still use the Walkley-Black method to analyze samples. This method should be avoided.[10] Quantifiers should confirm the process and equipment the labs will use before engaging them. The lab should use standard materials of known composition to calibrate instruments and should participate in interlaboratory comparison of results of analysis of reference materials.

Obtaining the dry weight of biomass samples requires drying them as soon as possible to avoid mold or loss of organic carbon from decomposition. If analysts cannot immediately dry field samples, they should be air-dried or, if that is not possible, refrigerated. Ideally, samples should be dried by cutting them into small pieces and desiccating. However, heat is often used instead of desiccation. Heat does not remove quite as much water from wood as can be removed by desiccating ground samples. For samples from live plants, drying should occur at 60°C to 80°C, as higher temperatures can volatilize modest amounts of the organic carbon. Drying should continue until the weight of the sample becomes constant, indicating that all the water has been driven out. This usually takes several days, and more time for segments of branches longer than a couple centimeters. Quantifiers should weigh the dried samples immediately before they reabsorb moisture, especially in humid climates.

Quantifiers must then analyze the proportion of the dried biomass that is carbon using the modified Dumas combustion method. This entails oxidizing a small sample at very high temperatures, typically about

1,000°C, and then using infrared gas absorption or gas chromatography to measure the amount of CO_2 emitted. This technique is extremely accurate and precise if samples are homogenized well (since only 10–20 mg is used for the analysis, obtaining a representative subsample is critical) and equipment is well calibrated. Other methods such as near-infrared reflectance (NIR) and nuclear magnetic resonance (NMR) provide accurate results, but the equipment and training needed to use them are not widely available.

Finding the Total Carbon Content of a Plot

To determine sequestration, quantifiers must convert plot measurements to carbon stock on each plot at each time, find the change on each plot over time, and scale up to the project area. The carbon stock on each plot is the sum of the stocks of all the carbon pools within the plot, such as live trees, other live plants, woody debris, and the forest floor. When field measurements are weights, such as the weight of litter collected from a subplot of a specified area, field measurements can be converted to carbon by multiplying them by the proportion of weight that is carbon, then scaling up.

When field work measures the sizes of things, these sizes must be converted to weight to calculate carbon stocks. A large part of calculating forest carbon sequestration is conversion of data about the sizes of trees and the frequency with which they occur into carbon mass. A key step in this process is calculating the mass of carbon in each measured tree.

The species, height, and diameter of a tree reliably relate to the mass of that tree. Individual trees of a given species and shape have similar sizes and shapes of trunks and branches and similar wood densities. There is some variation in the relationship of mass to height and diameter, however, depending on the variations within some species, climate, and (to a lesser degree) the conditions under which an individual tree grows. As a result, equations used to predict tree biomass as a function of height and diameter should, ideally, be created from trees in the region in which the equation will be used. Otherwise, they should at least be created from trees that grew under climatic conditions similar to the conditions where the equation will be used. Equations that predict tree weight or volume as a function of tree sizes are also called allometric equations. Quantifiers may use existing biomass equations or develop new ones if appropriate equations are not available.

Equations that predict carbon content of trees from height and diameter are available from a variety of sources. U.S. Forest Service publications contain a wealth of information, including biomass and volume equations for a wide range of species. Quantifiers may need to search the website of individual Forest Service research stations because system-wide searches do not seem to find all applicable materials. Many Forest Service research publications are available for free download.

BIOPAK, software the U.S. Forest Service offers at no charge, includes biomass equations for a variety of North American plants (see http://www.fs.fed.us/pnw/). BIOPAK provides references for the original sources of the equations, which can help users determine their applicability. However, although an extraordinary resource, BIOPAK is not easy to use, and most quantifiers will search for other equations to use in electronic spreadsheets or other programs. Clark et al. (1986) provide equations for eastern North American hardwood species, and Clark (1987) gives sources for equations that predict total aboveground biomass, or the mass of specific components, for southern U.S. tree species. Aldred and Alemdag (1988) provide sources for predicting total aboveground biomass of specific tree components of Canadian trees.[11]

The Internet or a forestry library can also be a source of biomass equations. A search that includes the name of the species, the word "biomass," and the words "equation or estimat*" will likely turn up references. (The * serves as a wildcard in most search programs, and it will return any word that starts with the letters preceding the wildcard, such as either "estimate" or "estimation.") If such a search does not yield results, the word "biomass" can be replaced with "volume" and the search repeated.

Stem-volume equations are available for many species because the volume of tree trunks is commercially important for the production of lumber and wood fiber products such as paper. Such equations use information on the density of carbon in different species to convert the volume of wood, as indicated by field measurements, to carbon mass. Some volume equations are for wood only; others include both wood and bark. Because the wood-products industry developed many of these equations, they often exclude branches, foliage, tops, and stumps, but quantifiers can estimate crown mass as a function of stem size or mass.

An equation may apply to a single species or group of species, or it may be limited to a single species grown under a specific management regime. Experts develop the equations by cutting and weighing trees and then using regression analysis to develop an equation that relates the measured weights to the physical characteristics of the trees. The resulting equations apply only to the range of tree sizes from which they were developed. Quantifiers are often tempted to use equations intended to estimate the biomass in smaller trees to estimate biomass in large trees if the equations match the species and location. However, that approach may cause significant errors, and there will be no way to detect them. If an equation for large trees is needed, it is better to adapt an equation for large trees of a similar species that have a similar growth form than to apply an equation for smaller trees. A biomass equation can be adapted for application to a different species having a similar growth form by adjusting for any difference in the specific gravities of the woods of the two species in question.

Equations for shrubs and very small trees often use the diameter measured at the base, just above the root crown, rather than the diameter at breast height. Some shrub equations use canopy diameter rather than stem diameter. Some equations provide volume rather than biomass. (Quantifiers can convert volume to biomass by multiplying by the density or specific gravity; see Appendix 10.) Some equations calculate the dry-weight biomass or carbon mass of a single tree, typically as a function of diameter or both diameter and height. Most such equations are made from measurements of the aboveground parts of trees. Some equations predict biomass of a single component of a tree, such as foliage, branches, bark, or bole wood. Relatively few studies of the root biomass of trees have been published, although general equations for North America predict root biomass as a function of aboveground biomass and diameter.

Biomass equations that do not use tree heights give less reliable estimates of biomass than equations that use both height and diameter. This is because tree height—for a given diameter—can vary tremendously as a function of site productivity and the tree density under which the stand developed. However, much of the time, using both height and diameter gives no more than 10% more accurate predictions of biomass than using diameter alone. If forest stands are managed and biomass equations are developed from similar stands and not applied to old-growth trees, equations that use only diameter and species to predict biomass should be adequate.

Quantifiers must specify the equations they will use to calculate biomass, and the properties of the trees that will be used to drive them, before designing the sampling system and specifying the field protocols. If quantifiers do not select equations until later, field crews may not collect all the information needed for the calculations, and the money spent on sampling may be wasted.[12]

If more accurate equations become available during the project, or if the factors used to drive the equations change, quantifiers may be able to adopt the new equations. However, this may not work if measurements from earlier fieldwork cannot drive the new equations. A project's monitoring plan may also call for developing new factors, such as site-specific densities of woody debris not present on the site earlier. Waiting until the second measurement of biomass stocks to develop new density factors or equations is often efficient, as other analysts may have created usable factors or the project's needs may have changed. All the needed data must be collected at the appropriate time, though, and quantifiers should use extreme caution in changing methods for collecting information because such changes may rule out comparisons of earlier and later biomass measurements.

If quantifiers do not have enough information to use a specific biomass equation for a given species, or if they cannot find an appropriate equation, they have several options. They can use equations developed for other species, they can create new equations, or they can use an equation for a group of species instead of an equation for a specific species (see Appendices 13–15).

Calculating Changes in Carbon Stocks

After calculating the mass of each carbon pool on a plot and scaling the results into common units (such as tons of carbon per hectare), quantifiers sum them to determine the total carbon mass for each plot. They then subtract the amount of carbon present on that plot at the start of the project from the new amount to find the change in carbon stocks. Of course, quantifiers must use the same biomass equations to estimate both amounts.

Using permanent plots allows finding the change in carbon stock on each plot before expanding to the change in carbon stock on the project area as a whole. This approach of finding the change on each plot is called paired plot analysis; the carbon density of each plot measured at a later time is paired with the carbon density on each of those plots measured at the start of the project. Paired plot analysis is different from the typical analysis of difference of means. The difference of means would be found by calculating the mean estimated carbon stock of the entire project area at the start of the project, calculating the mean estimated carbon stock of the entire project area at a later time, and then finding the difference between these two estimated mean carbon stocks. Paired plots are used because pairing plots through time reduces variability, thus giving a more precise estimate of the change in carbon stock.

After calculating the change in carbon stock on each plot, quantifiers then calculate the change in carbon stock for the entire project area, along with its uncertainty. If the project has only one stratum[13] and has installed plots randomly, then the overall change in carbon stock is the average of the changes in all the individual plots, in metric tons of carbon per acre. The average change per plot is

$$\Delta C_{avg} = \left[\sum_{1}^{i} \left(C2_i - C1_i \right) \right] / n \qquad \text{Equation 6.1}$$

where ΔC_{avg} is the average amount of carbon gained throughout the project area, $C1_i$ is the amount of carbon observed on subplot i in sampling site s at time 1, $C2_i$ is the amount of carbon observed on subplot i on plot s at time 2, and n is the number of subplots in the area sampled. If the project has multiple strata, this calculation is performed for each stratum.

Before applying Equation 6.1, quantifiers should convert $C2_i$ and $C1_i$ to tons of carbon per hectare so that ΔC_{avg} will also be in tons per hectare.

The next step is to calculate the mean estimated change in carbon stocks for the entire project area. If all plots have the same area, quantifiers can do this by calculating the average change per plot (in units of tons per hectare) and then multiplying by the total hectares in the project:

$$C_{seq} = \Delta C_{avg} \times A \qquad \text{Equation 6.2}$$

where C_{seq} is the total amount of carbon sequestered by the project (in tons), ΔC_{avg} is the average change in carbon stock observed on plots (in units of tons per hectare), and A is the total area of the project lands. If the project is stratified (see Appendix 1), C_{seq} is the sum of the amounts of sequestration calculated for each stratum.

As described in Chapter 2, the project's net GHG benefit is the overall gain in sequestration minus the baseline ($C_{seq} - B$) and inadvertent emissions from project activities (see Appendix 2). The project's offsets equal the net GHG benefit minus leakage and the statistical uncertainty in the calculations (see Appendix 3).

Of all biotic offset projects, forestry projects have the potential to provide some of the greatest GHG benefits—both per hectare and per dollar invested. Thus they can provide an important contribution to a carbon market. However, forestry projects are complex, and their benefits are difficult to measure precisely. Careful planning and strict adherence to the procedures outlined here is essential to the success of these projects.

Chapter 7

Step 4: Quantifying the Carbon Sequestered in Soil

Although soil and plant detritus contains 1.5 to 2 trillion metric tons of carbon worldwide, carbon accounts for only 1 to 5 percent of the soil on the surface and less than 1 percent of soil below the surface. Moreover, the amount of carbon a land-use project sequesters is usually small compared with the amount of carbon already stored in the soil. The gain is almost always less than 10 percent and often less than 5 percent, and if carbon is measured to a depth of only 1 meter, the gain is usually less than 3 percent.

These attributes make quantifying the offsets a soil project produces challenging. Measurements must be precise and designed to account for variations in soil carbon from one from location to another. This chapter provides an overview of how to design a quantification system to achieve those goals (see Appendix 16 for more information).

Because a system for quantifying soil carbon is complex, most developers will want to consider projects that encompass at least 25,000 acres and sequester at least 25,000 tons of carbon,[1] to make the costs of measuring changes in soil carbon cost-effective. Project developers with smaller land areas, or who are seeking to generate fewer tons of offsets, should consider aggregating their lands to reduce the cost per ton of measuring and verifying offsets. As a last resort, smaller projects may be able to rely on modeling to estimate how much carbon they sequester. However, some carbon markets or regulatory systems may not accept less rigorous quantification, or the resulting offsets may sell at a lower price.

In most cases, developers must quantify the carbon sequestered in soils on project lands by

- Designing a system for measuring changes in the amount of carbon in the soil.
- Taking field measurements of carbon stocks at the start of the project.
- Monitoring project conditions over time to assess whether managers have implemented changes in land-use practices and to gauge the amount of carbon stored.
- Remeasuring carbon stocks in soil and calculating changes in those stocks.

Designing the Measurement System

Quantifiers must be able to document and accurately quantify the sequestration that occurs on project soils. Without an acceptable method for estimating benefits, project developers cannot say with confidence that sequestration has occurred, and thus they will likely be unable to market their offsets.

A sampling design and protocol for analytic measurements must be designed at the outset to accurately quantify the changes in soil carbon over the project period. Sampling and analyzing soil samples can be costly, so the design of the sampling program can strongly affect the cost of the project and its profitability. The goal is to select a sampling program that achieves a level of precision high enough to detect tons of sequestered carbon without incurring untenable costs. An appropriate sampling design and analytical framework, careful fieldwork, and high-quality laboratory testing can provide a high level of precision for acceptable cost. For some types of projects, establishing an adequate sampling strategy may prove prohibitively difficult (see the sidebar on erosion).

Quantifying carbon sequestration is made especially difficult by the fact that the increase in soil carbon stocks in most projects will be less than 10 percent of the total soil carbon content—and much less if deeper soil is sampled. This means that if quantifiers need to measure the net sequestration (or change in the soil carbon content) to an accuracy of 10 percent (as recommended in Appendix 3), they will have to measure the total soil carbon content to an accuracy of at least 1 percent.

For example, consider a project that switches from plowing to no-till farming. Such projects will typically store on the order of 2 to 4 tons of carbon per acre. Suppose that the project actually stores an extra 4 tons of carbon per acre. Further suppose that the project developer hopes to get credit for sequestering at least 3.5 tons of carbon per acre. That means that the uncertainty in the measured change in carbon stock must be no more than 0.5 tons of carbon per acre (see Chapter 3).

Achieving that level of accuracy can be challenging. Project developers can increase their odds of meeting it by adopting a strategy that involves choosing sampling sites randomly to avoid bias, using paired sampling,[2] selecting enough sampling sites to ensure statistical accuracy, and adopting stratified sampling to further increase statistical power (see Appendix 1). A typical project would probably require about 50 to 100 sampling sites to achieve that level of statistical precision, with one site located every 100 hectares.[3] That means each field would probably include only one plot, and some fields would have none. With this wide a distribution of plots, projects need a system to ensure that the locations of sampling sites are in fact random. If developers are using a GIS program to map the project area, the software can randomly assign plot centers.[4] If developers are not using a GIS program, quantifiers can use a random-sampling technique to assign plot locations manually.

A sampling strategy should include a detailed protocol for collecting samples of a specified volume from numerous sites. Field crews will have to remove rocks and roots from each sample. The soil is then ground and mixed, and a subsample is analyzed in a lab to determine its carbon content. The sampling protocol should specify

- The number and spatial arrangement of soil samples to be taken at each site.
- The steps field crews should take if they cannot extract a sample at the specified location.
- The diameter of the soil cores and the depth to which field crews will collect each core.
- The guidelines for how crews should deal with materials on the surface of the soil and for how they should label, package, and handle samples.

Coring is the most efficient soil-sampling technique that uses commercially available tools. In this approach, field crews collect soil cores from specified sites by hand or by using hydraulically powered coring

The Challenges of Erosion-Abatement Projects

Conservation practices such as contour plowing, planting of grass strips, and reduced tillage can greatly reduce erosion and thus increase the amount of carbon in soils. However, reducing erosion may merely prevent the transport of stored carbon off project lands, rather than increase the total amount of carbon stored inside and outside the project. Moreover, carbon stocks under a given type of vegetation for a particular soil and climate tend to approach equilibrium. Thus when erosion removes carbon from a site, the vegetation will usually store more carbon for a period of time to make up the deficit. If soils outside project boundaries trap the eroded carbon, and vegetation begins sequestering more carbon at the eroded site, erosion may actually increase rather than decrease overall carbon stocks (Smith 2005).

machines. (The latter can take larger and deeper cores, but the cores must be transported by a truck or tractor.) Obtaining the desired level of precision requires mixing multiple cores from each site to account for variability and reduce measurement error.

Corers employed to sample soil are usually tubular and range in diameter from about 2 to 8 centimeters. Although using the smallest-diameter corer that will gather intact samples is most cost-effective, a larger corer may work best in soils with some buried gravel. If crews are uncertain about whether a particular corer will collect samples to the desired depth, it is cheaper to field-test the corer than to choose a large-diameter corer, transport hundreds of kilograms of soil, and spend days processing the larger samples.[5]

Coring may not work if crews are taking measurements at multiple depths in soils that compact a great deal when cored, that contain large numbers of rocks or buried wood such as roots, or that are very noncohesive (such as dry sand). If the amounts of rock or buried wood are so great that crews cannot extract cores after a few attempts, crews may need an instrument designed for sampling noncohesive materials, such as a bucket auger for sampling sand. A drawback of bucket augers is that they extract disturbed material, not an intact core, thus mixing soil from a range of depths.

Quantifiers may soon be able to use new portable technologies such as laser-induced breakdown spectroscopy, inelastic neutron scattering, and near-infrared spectroscopy to measure carbon content in the field. However, these emerging technologies require further testing and refinement before they become accepted approaches to measuring changes in soil carbon.

Deciding on Sampling Depth

The decision of how deeply to sample soil is perhaps the most important decision in designing a system for measuring soil carbon. Most of the increase in carbon in soil projects will usually occur in the top few centimeters of soil. However, these increases may simply represent carbon redistributed from deeper depths, with the project having produced little or no net sequestration, especially in the first few years after a change in land management or vegetation. This may be the case, for example, during the first decade after a switch from plowing to no-till farming because the lack of plowing slows the transport of plant material (and its attendant carbon) to lower depths. Such projects may even see an overall loss of soil carbon during the first few years, especially in dry climates. For this reason, no-till sequestration projects should usually sample at least the entire plow layer of soil, which typically extends about 20 centimeters below the surface.

On the other hand, sampling deeper than 20 to 30 centimeters may not be worthwhile unless species and soils have unusually large amounts of root mass or carbon deposition at greater depths. Projects should conduct deeper sampling if amounts of soil carbon may decline at those depths, or if the project will establish deep-rooting grasses, which can add significant amounts of carbon to soil to 2 meters, and small amounts to 4 meters. Sampling to deeper depths makes discerning sequestration against a larger volume of soil more difficult. Developers may choose to forego measuring some of the carbon gain at those depths if the cost of doing so is greater than the value of the carbon or if attempting to measure some of the carbon stock would dilute the precision of the overall measurement. However, if there is serious concern that the change in land management will cause loss of carbon deeper in the soil, sampling must encompass the depth where loss may occur.

Determining the Number of Cores

A detailed measurement plan must specify techniques for establishing permanent sampling sites where crews collect a set number of cores from each site with a specific spatial distribution, which are then mixed into a single sample and sent to a laboratory for analysis. Establishing permanent plots allows crews to return years later to remeasure soil carbon and calculate the change on each plot, which gives the overall results for the project statistical power. To help field technicians find each plot during later sampling periods, crews should mark each plot center, such as by placing an electronic marker in the soil. (An electronic marker is an antenna

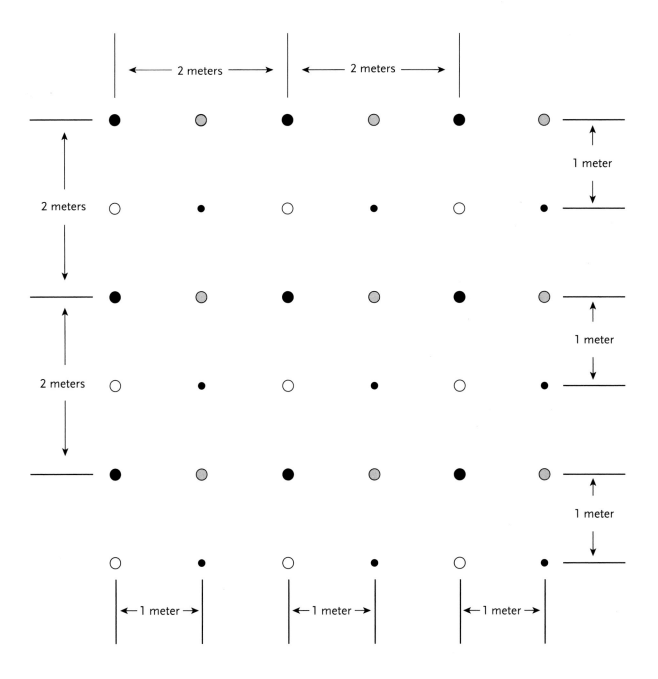

2 meters   2 meters

1 meter

2 meters

2 meters

1 meter

1 meter

1 meter   1 meter   1 meter

buried magnetic markers

● initial sampling

◯ first resampling

○ second resampling

• third resampling

Figure 7.1 Layout of cores at a sampling site. Establishing permanent plots allows crews to return years later to re-measure soil carbon and calculate the change on each plot, which gives statistical power to the the overall results for the project. Although most aspects of sampling should remain constant from one measurement round to the next, crews should extract soil cores at points displaced from those used during prior sampling to ensure that the results are not influenced by disturbances incurred by the sampling itself.

that is encased in plastic [to keep it from rusting] that is buried deep enough so any likely disturbance, such as plowing, will not move it.) Field crews collecting later samples use a electronic locator, which is similar to a metal detector, to find the marker.

Plots are typically at least 2 by 5 meters (10 square meters), but not larger than 9 by 9 meters. Field crews collect a predetermined number of cores around the plot center, even though the center may be on the corner of the grid (see Figure 7.1). The spacing between cores, usually 2 meters, should not be so great that plots cross soil types or landforms or that plots vary significantly in some other way.

The number of cores to collect and mix on each sampling site is a key determinant of whether the sample accurately represents the amount of carbon in soil. Collecting and mixing as few as six cores per plot may work for soil that has been tilled many times, as tilling makes soil more homogeneous. Mixing 10 to 16 cores per site is best for soil that has not been tilled or for soil where woody plants have been growing.

If early measurements show that collecting fewer cores per site or using a smaller-diameter corer will yield measurements of acceptable variability, quantifiers can make those changes to reduce costs. Quantifiers may also decide to analyze different depth increments separately and to collect fewer cores to the full depth. For example, if the plow layer is 20 centimeters deep and crews sample to a 50-centimeter depth, quantifiers might analyze the 0–20-centimeter layer separately from the 20–50-centimeter layer. This approach can reveal where sequestration is or is not rising. However, it does not give more statistical power because it does not increase the number of plots. Measuring depth increments separately also increases the costs of transporting and processing soil samples. Collecting fewer deep cores at each site and processing an additional sample at the added depth is usually more cost-effective.

Although most aspects of sampling should remain constant from one measurement round to the next, two aspects *should* change. Crews should extract soil cores at points displaced from those used during prior sampling to ensure that the results are not influenced by disturbances incurred by the sampling itself (see Figure 7.1). Consider a sampling design that removes nine cores from each sampling site in a 4-by-4-meter grid, with 2 meters between intersections. During initial sampling, the northwesternmost sampling point is the reference point. For the next round of sampling, crews could displace each sampling point 1 meter south. During a third round, they could displace each sampling point 1 meter east of the initial points, and during a fourth round they could displace each point 1 meter south and 1 meter east.

The second aspect of measurement that should change is the location of points for sampling decomposed organic material. If a layer of such material sits above mineral soil, sampling will reduce the carbon stock in this layer for at least several years, and possibly for more than a century. To keep this local disturbance from biasing the estimate for the project, field crews should displace locations for sampling decomposed organic materials from earlier locations. However, crews can sample such litter at the same locations if they do so only every five years or more, as most litter will have accumulated during that time.

A project's sampling protocol should specify how to deal with obstacles such as large rocks and trees that prevent crews from collecting cores at designated points. If they cannot extract a core from a specified point, moving a fixed distance such as 10 centimeters north may introduce less bias than moving only as far as needed. If crews still cannot extract a core, they can move another 10 centimeters north and try again. If they hit bedrock, they should collect a sample to that depth and record it. If no soil is present, they can also record that fact.

Quantifiers should remove plots from the measurement system only if the land on which they sit has been dropped from the project. Plots should not be removed because they have been bulldozed or otherwise disturbed, a road has been built through them, or a river has shifted its channel. If field crews fail to find a plot marker after a diligent search, they can consider the plot lost and establish a new one at the prescribed location.

Determining Frequency of Measurement

The optimal interval for measuring changes in soil carbon depends on the rate of change, the cost of conducting measurements, and the value of any offsets. More frequent measurements reveal any shortfall in carbon sequestration quickly, giving developers a chance to address problems. Project developers also wish to deliver offsets and get paid for them as soon as possible. In addition, uncertainty rises as time passes since the last measurement, eroding the amount of offsets verifiers will accept and lowering the price these offsets might command. On the other hand, lengthening the time between remeasurements spreads quantification costs over a larger amount of offsets, which tends to increase the profitability of the project. The challenge is to balance the tension between delaying remeasurement to reduce costs and hurrying it to verify sequestration as it occurs.

Changes in soil carbon are typically not measurable from one year to the next because the change is too small relative to the total carbon stock. Moreover, such frequent measurements may prove unreliable because sequestration varies from year to year depending on the weather. The dynamics of soil carbon during the first one to three years after a switch from plowing to no-till farming are also poorly understood, and it is unclear how quickly net sequestration begins. Given current technology, costs, and annual variability in sequestration, most project developers should probably choose to measure soil carbon every five to 10 years for the first 10 to 15 years. Developers may plan further measurements in later years, or they may simply monitor a project to ensure that conditions are conducive to maintaining sequestration, if they expect the soil to store little additional carbon. Projects usually measure soil carbon shortly before they end, to determine whether they have met their overall target.

A project's measurement plan may call for a hybrid approach wherein quantifiers measure carbon stocks for several years and then use the resulting data in models to calculate changes in later years. Before choosing a hybrid approach, project developers should assure themselves that the needed modeling capacity is available at an acceptable cost. Although leading soil carbon models are available free of charge, paying people with the expertise to run them may prove costly.

Designing a sampling system; conducting an initial measurement of soil carbon within the project area; measuring changes in soil carbon later; and paying for laboratory costs, data analysis, and verification can easily cost several tens of thousands of dollars. After calculating the number of plots needed and setting a schedule for remeasurement, developers may wish to estimate the cost of all the quantification work over the lifetime of the project to see if it is likely to detect enough sequestration to be financially viable.

Developers may choose to monitor at more frequent intervals to determine whether land managers have implemented the promised activities and those activities are yielding the anticipated sequestration rates. These extra monitoring activities should specify performance thresholds, such that if the thresholds are met, the project is likely to be sequestering carbon according to plan. A near miss might trigger further measurements to better understand project conditions, whereas a complete miss could trigger remedial action.

Thresholds may be quantitative rather than categorical. Suppose a project plans to boost soil carbon by increasing crop residue left on fields to 5 tons per acre. Field crews weigh the residue on small, randomly located plots. If the average mass is less than 5 tons per acre, or if more than 10 percent of the plots have less than 4 tons per acre, such a finding would trigger more intensive measurements of residue and modeling of the sequestration likely to result.

Quantifying Carbon in Samples

The most common techniques for analyzing the proportion of soil that is carbon are based on measurements of the emissions from the dry combustion of soil samples. (This approach is quite similar to that described in Chapter 6 for analyzing samples collected in a forestry project.) Cores of a known volume are collected, dried, and weighed. The weight is then divided by the volume to yield soil bulk density.

To find the amount of carbon in the sample, labo-

Modeling Future Changes in Carbon Stocks

Developers typically use modeling or extrapolation from benchmark sites to estimate how much sequestration a project will produce before they embark on it. However, developers may also use data collected during the initial measurement of carbon stocks to model potential sequestration and to check progress during the project.

Developers need at least one modeling run for each combination of conditions. For example, if the project encompasses two different soil textures and cropping regimes, they need to run the model for each combination of soil type and cropping regime. Modeling is typically done on a per-hectare or per-acre basis and scaled up. Two user-friendly computer programs, the soil carbon tool of the Intergovernmental Panel on Climate Change (IPCC) and the COMET model (both available free online), quickly give a scientifically based estimate of changes in soil carbon resulting from changes in land management. A third soil carbon model, CENTURY, can make site-specific predictions based on data from land managers, an initial measurement of soil carbon, and other sources. CENTURY has been widely validated and is also available online at no cost. However, formatting data for use in this model, selecting factors for the calculations, and assessing outputs requires substantial expertise. The information needed to operate the IPCC and COMET models includes soil texture, cropping regime, tillage practices, productivity, and nutrient inputs, whereas the CENTURY model also requires historic weather data from a nearby location.

Assessing Uncertainty

Regardless of whether quantifiers use measurements or models to determine changes in soil carbon, they must assess the uncertainty in the calculated offsets. Using site-specific information to better represent actual carbon dynamics may yield more precise estimates, reducing uncertainty. Smaller uncertainty ranges, in turn, may allow quantifiers to detect more of the sequestered carbon with a high level of con-

fidence, thus producing more credited offsets and gaining a higher price for the offsets.

Empirical measures of uncertainty are far better than expert opinion. Studies have shown that people often think their predictions are much more accurate than they turn out to be.[6] Whereas an evaluation of uncertainty based on actual measurements of soil carbon stocks is fairly straightforward (see Appendix 3 on statistics), such an evaluation based on models is more problematic. One approach to quantifying the uncertainty of estimates by a model involves finding the difference between modeled and observed outcomes in a number of cases and using that difference to calculate the standard deviation of the model's errors.

Some analysts use Monte Carlo analysis to estimate uncertainty. Properly done, Monte Carlo analysis examines variation in predicted outputs from thousands of model runs, where the inputs for each run are randomly selected from the possible range for each input.[7] For example, suppose that a model uses the amount of rainfall occurring each month as an input, and the model is run with rainfall records for a 25-year period. For each month, there are 25 possible values for the amount of rainfall for that month. During each run of the model, for each month, the Monte Carlo analysis would randomly select a year and use that amount in the model run.

Using Monte Carlo modeling to estimate uncertainty assumes that the model correctly represents dynamics in the physical world. This assumption is never totally correct; all models, by definition, are simplifications. If the model represents the world reasonably accurately, the modeled uncertainty will be close to the observed uncertainty in the world. If the model does not reliably depict the world, the modeled uncertainty may be much smaller or larger than the true uncertainty. Monte Carlo simulation is appropriate for a complex model such as CENTURY. The IPCC's soil carbon tool and COMET do not allow enough variation in inputs for users to perform Monte Carlo simulations. However, the COMET tool does provide estimates of uncertainty by comparing differences between modeled outputs and measurements at benchmark sites.

Validating Model Estimates

If a project will run for a long time and quantifiers will calculate soil carbon stocks more than twice, modeling can be very useful in determining whether initial projections match what is occurring. Initial measurements can be used as inputs to model runs, and predicted soil carbon values can be compared with those observed during the second field measurement. If the modeled and measured values match, users can have much higher confidence in model projections of later sequestration. If modeled and measured sequestration amounts do not match, project developers can adjust projections of future sequestration. Only a few sampling points, spanning the range of conditions across the project area, need to be measured during the second field measurement. Quantifiers can run the model using information from these sites as a check on the reliability of predictions for all sites.

Sensitivity analysis can be used to determine which inputs have the greatest impact on outputs. Quantifiers can then focus on obtaining more reliable data for those input variables.

ratory analysts first take a small subsample from each core and measure its mass. They then oxidize (or burn) the subsample at a very high temperature, using infrared gas absorption or gas chromatography to measure the amount of CO_2 emitted. Analysts can convert this amount to grams of carbon by dividing it by 3.667 (the ratio of the mass of CO_2 to carbon). They can then find the amount of carbon in soil per unit of area by dividing this quantity by the mass of the subsample and multiplying it by the bulk density of the sample and depth of the core (see Appendix 16). The amount of carbon sequestered in soil is best expressed in tons per hectare.

This technique is extremely accurate if samples are prepared properly and equipment is calibrated and used correctly.[8] Crews must be careful to collect all soil from sample cores and to exclude soil that is not from the cores. If samples will not be processed for several days, they should be refrigerated or frozen to slow decomposition and loss of carbon.

To obtain an accurate reading of soil carbon, laboratory staff should thoroughly mix the entire soil sample or preferably mill the entire sample except for roots or other materials that are not classified as soil. At minimum, it is essential to mill a subsample of soil to a very fine texture and homogenize it. If such preparation is insufficient, carbon numbers will be highly variable, and quantifiers will not detect the modest amounts of carbon that projects are likely to sequester. (Subsamples typically weigh only a fraction of a gram, although their weight may vary with their carbon content.)

If a significant proportion of the particles in the soil are larger than 2 millimeters, analysts should grind a sample of this material and test it for the presence of carbon. If they find carbon, they should process 10 to 30 samples to see if such material contains a uniform percentage. If it does, they can use that percentage in evaluating the overall amount of carbon that such material contributes to soil samples. If the carbon content in this material varies significantly, analysts should measure more samples until they find an acceptably small standard of deviation. Porous rocks such as sandstone and some volcanic rocks are particularly likely to include carbon. Rocks with carbonates, such as limestone, include inorganic carbon that will produce CO_2 when combusted, so their presence would require further analysis to distinguish organic from inorganic carbon.

A number of universities operate high-quality analytical facilities and will analyze the amount of carbon in soil subsamples for a modest fee. A useful approach is to rely on a lab that analyzes samples jointly with other labs and compares results. After chemical analysis of soil subsamples, quantifiers should archive remaining samples for reanalysis later, if necessary.

Determining the Change in Carbon Stocks

One might assume that the change in carbon stocks at any specific site is simply the difference between the mass of carbon per unit of area at the beginning of the measurement period and the mass at the end. However, if the bulk density of the soil changes over time, the calculation process must account for this change. Failure to do so can lead to errors that range from doubling actual sequestration to falsely concluding that the soil has lost carbon when it has gained carbon (Gifford 2003).

Changes in bulk soil density usually reflect the fact that soil has become more or less compacted. For example, soil density usually rises for several years after land managers switch from plowing to no-till farming. That is because the soil collapses until soil aggregates form and re-create the porous structure found in productive soils with little disturbance. The height of the soil surface usually changes along with bulk soil density: when soil compacts, the surface drops; when soil becomes less compact, the surface rises.

When soil density increases, resampling to a given depth captures more soil. The inverse is also true: if soil density decreases, resampling to a given depth captures less soil. For example, suppose that in project year 1, crews sample soil to a depth of 20 centimeters. Further suppose that over the next few years, the soil increases in density (or compacts) by 10 percent, and the surface drops. If resampling in year 10 also occurs to a depth of 20 centimeters, it will capture about as much soil as sampling to 22 centimeters would have captured in year 1 (see Figure 7.2).

To account for this effect, quantifiers must calculate bulk soil density for each sampling site each time they measure carbon stocks. They can do so by separating any rocks, roots, and other material larger than a specified size (such as 2 millimeters) from fine soil and then consulting soil-sampling manuals on how to measure the density of this material. This approach accounts for the fact that samples taken at different times may include more or fewer rock fragments and roots. (Unbiased measurement of the density of rocky soils requires the use of more laborious pit sampling, as corers cannot encompass large rocks and usually do not yield intact cores when encountering them.)

To determine whether soil density has changed during out-year sampling, field crews should extract an extra 5-centimeter portion of soil from the first few sites (see Figure 7.3). For example, if the initial sampling included soil to a depth of 20 centimeters, crews should remove soil from a depth of 20 to 25 centimeters as a separate sample.

Quantifiers then measure the density of several soil samples taken at the original depth. If densities are within 1 to 2 percent of remaining constant over time, field crews may stop collecting the extra depth increments. However, they should not discard the samples collected until the overall analysis is complete. If bulk density has changed, the change might be a fairly constant percentage across sites, or it might occur only under some conditions. Quantifiers may need to analyze 20 to 30 sites to discern a pattern. If they cannot detect a pattern, crews should collect the extra depth increment at all sites. Quantifiers then use those depth increments to correct for changes in bulk soil density (see Appendix 16).

After calculating the change in carbon stock at each sampling site and correcting for changes in bulk density, quantifiers then calculate the change in carbon stock for each plot. Next, they calculate the mean change in carbon per plot for all the plots analyzed. If the project has only one stratum[9] and has installed sampling sites randomly, then the average change in carbon per plot is

$$\Delta C_{avg} = \Sigma(C2_i - C1_i) / n \qquad \text{Equation 7.3}$$

where ΔC_{avg} is the average amount of carbon gained, $C1_i$ is the amount of carbon observed on plot i at time 1, $C2_i$ is the amount of carbon observed on plot i at time 2, and n is the number of plots. ΔC_{avg} will be in the same units as $C2_i$ and $C1_i$. As noted, quantifiers should convert plot measurements to tons of carbon per hectare before performing this calculation.

The mean estimated change in carbon stock is the average of the changes measured at each sampling plot (in metric tons of carbon per hectare) times the number of hectares in the project:

$$C_{seq} = \Delta C_{avg} \times A \qquad \text{Equation 7. 4}$$

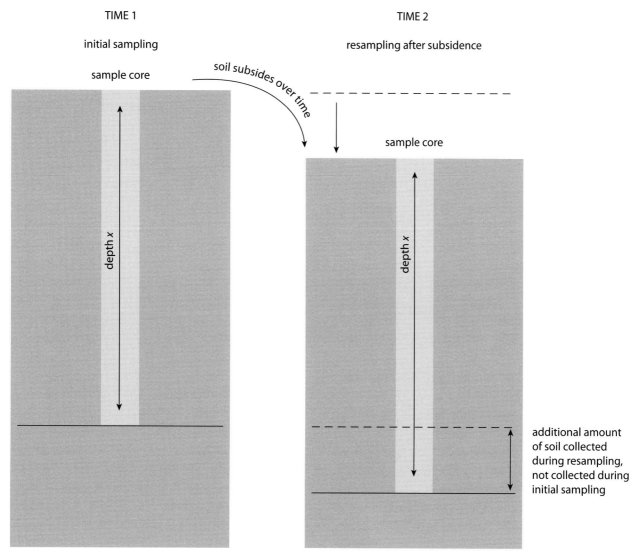

TIME 1

initial sampling

sample core

soil subsides over time

TIME 2

resampling after subsidence

sample core

depth x

depth x

additional amount
of soil collected
during resampling,
not collected during
initial sampling

Figure 7.2 The effect of changes in bulk soil density on the amount of soil sampled. Soil density can change over time (for example because of compaction and subsidence) and often changes in density are accompanied by changes in soil height. When soil density increases (as illustrated here), resampling to a given depth captures more soil. The inverse is also true: if soil density decreases, resampling to a given depth captures less soil. To account for this effect, quantifiers must calculate bulk soil density each time they measure carbon stocks.

where C_{seq} is the calculated amount of carbon sequestered over the project lands in metric tons of carbon, ΔC_{avg} is the average of the changes per unit of area measured at all plots (as metric tons per hectare), and A is the number of hectares encompassed by the project. Note that the mean estimated change is not the amount of offsets credited to the project. The credited offsets are equal to C_{seq} minus inadvertent emissions, the baseline, the uncertainty or confidence interval, and the leakage.[10]

Calculations of total project sequestration are somewhat more complex if the project is stratified. In that case, project sequestration is the sum of the amounts of sequestration calculated for each stratum. Quanti-

sample core

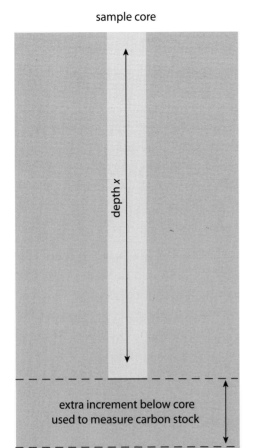

depth x

extra increment below core
used to measure carbon stock

Figure 7.3 Extra sampling to calculate soil bulk density,
Soil density can be obtained from the same cores used
to measure carbon content by extracting an extra
5-centimeter portion of soil from the first few sites.

fiers calculate sequestration for each stratum using the
method for a project without stratification.

Soil projects offer an opportunity for farmers and land
managers to participate in burgeoning carbon markets by making only minor adjustments to their normal practices, such as by switching to no-till farming.
With a moderate investment in labor and monitoring
equipment, landowners can realize extra profits while
taking steps to absorb greenhouse pollutants from the
atmosphere.

Chapter 8

Step 5: Quantifying Greenhouse Gas Emissions from Manure

A manure offset project usually entails moving from a manure-handling system that releases a large amount of methane (most likely an open anaerobic lagoon, but perhaps a liquid slurry, deep litter, pit, or anaerobic digester) to a system engineered to capture and burn methane. Although burning methane produces CO_2 emissions, such a project can have a significant GHG benefit because methane is about 20 times more potent than CO_2 as a greenhouse warmer (see Chapter 2). Moreover, if the captured methane is burned in an electric generator, it can reduce the project's use of fossil fuels and may also create additional offsets.[1]

This chapter presents an overview of the protocols, analytic procedures, and calculations developers should follow in developing and implementing manure-based offset projects. (For more detail, see Appendices 17 and 18.)

Manure-Handling Systems

A wide range of systems for managing manure is available. The Intergovernmental Panel on Climate Change (IPCC) uses the following nomenclature to classify them (IPCC 2000):

- *Pasture*: Manure from pastured or range animals left where deposited and not managed.
- *Daily spread*: Manure collected daily from barns and spread on fields.
- *Deep litter* (cattle and swine): Dung and urine left to accumulate in stalls for long periods of time.

- *Poultry manure*: Manure collected in cages, with or without bedding.
- *Solid storage*: Dung and urine collected from stalls and stored for months, with or without drainage of liquid, followed by some other use or disposal method.
- *Dry lot*: In dry climates, litter allowed to dry in stalls before removal.
- *Pit storage*: Combined dung and urine stored in vented pits below stalls.
- *Liquid slurry*: Dung and urine transported and stored for months in liquid form, with water added as needed for handling, in tanks open to the atmosphere.
- *Anaerobic lagoon*: Dung and urine flushed with water into open ponds, where they are stored for more than a month, with or without the capture of methane or use of the water remaining after solids settle.
- *Anaerobic digester*: Management of slurry of dung and urine designed to facilitate conversion of solids to methane, which may involve temperature control, mixing, or pH management; the resulting gas may be released, flared, or used to generate power.
- *Composting-intensive*: Waste placed in a vessel or tunnel with forced aeration.
- *Composting-extensive*: Waste collected, piled, and regularly turned for aeration.
- *Aerobic treatment*: Waste collected as liquid and managed with forced aeration to provide nitrifica-

tion and denitrification; treatment produces large amounts of sludge, and total methane emissions depend on management of resulting sludge.

Anaerobic digestion projects are especially promising. By storing and processing liquid manure under low-oxygen conditions, anaerobic digesters convert much of the carbon in manure to methane. In fact, many manure digesters are designed to maximize methane production, bolstered by warmer temperatures and longer residency times. If this methane is released, it is a potent greenhouse gas, but a project that captures and burns the methane avoids these emissions and thus can create offsets. The methane can be flared in an open burner, or it can be burned in an electric generator or other equipment. A properly managed anaerobic digester can also avoid the release of ammonia, which becomes an air pollutant in high concentrations.

Overview of the Approach

Because decomposing manure can produce methane and nitrous oxide emissions,[2] quantifiers must account for both when calculating the greenhouse impact of a project based on changes in manure management. However, the methane emissions are usually much greater than nitrous oxide emissions. In addition, even though nitrous oxide has a much greater warming effect per unit of gas (see Chapter 2), nitrous oxide emissions typically have a minimal overall greenhouse impact compared with the methane emissions from manure.[3] Quantifiers should therefore focus on accurately measuring methane emissions; they can often rely on published tables to estimate nitrous oxide emissions.

Quantifiers must also calculate emissions of CO_2 because a project's offsets are based on the warming effect of the methane minus the warming effect of the CO_2 into which the methane is converted.[4] Quantifying emissions from manure projects therefore entails several steps:

1. Determining *project-specific ratios* used for converting rates of manure inputs to emissions of methane and nitrous oxide.
2. Making direct measurements at specified inter-

vals to determine the total wet volume or weight of manure handled in the project.
3. Calculating methane and nitrous oxide emissions as a function of manure input amounts during each step in the manure-handling system.
4. Summing emissions from all steps and converting them to CO_2e, for each accounting period.
5. This whole process needs to be repeated whenever the digestibility of animal feed, the carbon and nitrogen in the feed, the type of animals, or the distribution of their ages have changed enough to warrant it.

Determining Project-Specific Ratios

Project-specific ratios include:

- The dry mass of manure, as a function of wet mass or wet volume.
- The nitrogen content of the manure, as a function of manure mass or volume.
- Emissions of methane and nitrous oxide, as a function of manure mass or volume.
- Mass transferred to any later manure-handling process, as a function of inputs to the earlier process.

Quantifiers should measure these factors for each step in the manure-handling system that might release more than a tiny amount of emissions. For example, if a project holds digested waste in a lagoon until applying it to fields in the spring, quantifiers should calculate emissions from the digester, lagoon, and field application, as well as from those practices on comparison lands. If a project holds manure in a pit before periodically transferring it to a digester, quantifiers should count emissions from both the pit and the digester. Projects might also store manure as slurry before moving it in batches to an anaerobic digester and then applying the material remaining after digestion to fields. Projects might reuse some liquid to flush animal pens, use other liquid as fertilizer, and feed solids into a digester. If manure inputs or handling processes change significantly, quantifiers will need to develop new project-specific ratios.

Dry-Mass Ratio

If measuring project-specific emissions per unit of manure input is not possible, some carbon offset systems may allow quantifiers to use ratios observed at other locations. Emissions estimates based on ratios from other locations usually calculate emissions as a function of the dry mass of manure input. However, manure inputs at the project site are almost always measured when the manure is wet, expressed in volume (for example, by recording the change in depth of a holding tank) or weight (for example, by weighing the trucks transporting the manure). In these cases, a project-specific ratio must be measured for the dry mass of manure as a function of wet mass or wet volume.

Determining the dry-mass ratio requires collecting wet samples that span the range of manure types that occur in the project. The weight or volume of the samples is measured, the samples are dried, and then the dry mass is measured. The ratio is then obtained by dividing the latter value by the former (see Appendix 17).

Nitrogen-Content Ratio

Nitrous oxide emissions are typically estimated as emissions per total mass of nitrogen in the manure. Thus quantifiers need to measure the nitrogen content of the manure. This is done using procedures similar to those for obtaining dry-mass ratios: collecting samples, quantifying their weight or volume, drying the samples, and using standard analytical chemical measurements to determine the nitrogen content.

Emissions Ratios

Measuring emissions per unit of mass or volume requires sampling gases carefully and then using diffusion equations to find emissions as a function of gas concentrations. Quantifiers should measure methane and nitrous oxide emissions over a period of several weeks or months to determine the rate per unit of wet manure input at each step in the manure-handling system.

Sampling methane and nitrous oxide emissions from open manure-handling systems is difficult. Emissions from manure spread on lands vary greatly, with a substantial portion of annual emissions occurring within a day or two after soil becomes saturated with water. Even when gas production is more constant, such as from a manure lagoon, changes in temperature produce variations that complicate sampling and calculations. Quantifiers assessing emissions from open systems should therefore work with technicians experienced in taking these measurements.

If a project captures and meters the volume of gas from manure, this metering may replace measurements of manure volume or mass and the factors needed to calculate diffusion rates for the gas. Quantifiers can calculate methane and nitrous oxide production by multiplying the metered gas volume by the proportion that is methane or nitrous oxide. To do this, quantifiers must develop project-specific ratios for concentrations of methane and nitrous oxide in the gas by taking samples to a laboratory for analysis. (This assumes that the average chemical content of the gas remains close to the measured values. If it does not, quantifiers must perform periodic sampling and analysis.) Total methane production is the volume of gas produced times the proportion that is methane times the density of the methane (the default density of methane is 0.67 kilograms per cubic meter).

The warming effect of nitrous oxide produced by an anaerobic digester should not exceed uncertainty in the estimates of the warming impact of methane emissions. For this reason, quantifiers usually can rely on default ratios such as those published by the IPCC in 2000 and listed in Table 8.1.[5]

If gas production from a digester is metered, quantifiers must measure or estimate methane emissions as the proportion of methane that burning does not destroy because of incomplete combustion or fugitive emissions. An engineer can calculate the completeness of combustion as a function of burner design and use. In the absence of a site-specific measurement or calculation by an engineer, quantifiers can assume that a homemade burner combusts 95 percent of the methane, whereas an engineered burner combusts 97 to 99 percent. The IPCC default range for fugitive emissions from anaerobic digesters is 5 to 15 percent.[6] If quantifi-

Table 8.1 Estimated Rates of Nitrous Oxide Production as a Function of Mass of Nitrogen Input, by Manure-Management Practice

System	Mg N_2O/Mg N input	Comments
Pasture	0.031	
Daily spread	0.000	
Deep litter	0.008	Cattle & swine, < 1 month accumulation
Deep litter	0.031	Cattle & swine, > 1 month accumulation
Poultry	0.031	With bedding
Poultry	0.008	Without bedding
Solid storage	0.031	> 20% dry matter content
Dry lot	0.031	
Pit	0.002	
Liquid slurry	0.002	
Anaerobic lagoon	0.002	
Anaerobic digester	0.002	
Composting-extensive	0.031	
Composting-intensive	0.031	
Aerobic	0.002	

Notes: Mg = megagrams, or metric tons. This table reports emissions rates in terms of Mg N_2O/Mg N input, whereas the IPCC reports them as kg N_2O-N/kg N input. The ratio of mass of N_2O to N_2O-N is found by calculating the molar weight of N_2O relative to N_2, which is 44/28. Thus 1 Mg of N_2O-N yields approximately 1.57 Mg N_2O.
Source: Calculated from IPCC (2000).

ers cannot measure methane production (or emissions) from a digester, they can estimate it as a function of the amount and type of manure inputs and the duration and temperature of digestion.[7]

Seasonality can be a concern when developing ratios, as the temperature and the period over which manure is held affect emissions. Quantifiers should take measurements over an entire production cycle to find total annual emissions. For example, they might measure slurry temperature every month for a year to determine how it varies with air temperature. If quantifiers cannot make measurements over an entire cycle, they must model variations that may occur.

Quantifying Manure Inputs

Quantifiers may use any of several methods to find the mass or volume of manure added to the handling system. They should choose the method that provides adequate reliability for the least amount of work, given the project's equipment and layout.

For example, if manure is treated in tanks, quantifiers can calculate the volume of manure as a function of its depth. (They should subtract the residual amount left in the tank before refilling, using a spreadsheet program to record the annual tally.) Dry manure and sludge removed from liquid-processing systems are often transported for spreading or other disposal. As with tanks, quantifiers can calculate the volume of transport containers and determine the volume of manure by observing the depth of each load.

For continuous liquid flow systems, quantifiers need information on the flow rate. If a project is large enough, developers may be able to justify the cost of installing a meter where flows from different sources join together. Quantifiers may also estimate flows by multiplying the specified pumping rate by the amount of time pumps operate. If they use that approach, they should check the flow rate by diverting the pumped flow to a large container of known volume and measuring the length of time it takes for the pump to fill it. This observed pumping rate can be used to calculate the volume of material handled.

Quantifiers should measure flow as early in the treatment process as possible, before fractions are sepa-

Alternative Methods for Quantifying Manure Emissions

A small offset project may find the sampling and analysis required to determine project-specific ratios too costly. If measuring manure weight or volume would be very difficult or expensive, quantifiers can estimate manure emissions from feed inputs, although that approach is less reliable.[8] Efforts are also under way to develop models that can provide reliable, site-specific estimates of emissions.

The U.S. Environmental Protection Agency's *Inventory of U.S. Greenhouse Gas Emissions and Sinks, 1990–1991* estimates manure production for dairy cattle and the most common livestock in each state. Developers could use this information to quantify their emissions. However, developers who do not weigh their animals would have to estimate their weights as a function of livestock type, age class, and location.

The IPCC provides a method for estimating manure mass as a function of animal numbers, types, and location, with animal type and location serving as surrogates for information on animal diet and condition and air temperature (IPCC 1996, IPCC 2000). The IPCC's revised *Guidelines for National Greenhouse Gas Inventories: Reference Manual* also includes annual methane emissions for different types of animals in different parts of the world. However, many of the IPCC's weight estimates and emission ratios are based on expert consensus rather than surveys, and the organization does not quantify the reliability of these numbers. Farmers often claim that animals' weight gain is predictable in controlled operations, but they can reduce uncertainty by gathering more site-specific information.

rated or mass is lost through processing. If a water-supply hose is dedicated to flushing manure, a flow meter placed on the hose can record cumulative flow if checked periodically, such as monthly. However, water flow may give a poor estimate of the amount of manure solids because different workers may use different amounts of water to flush the same amount of manure. If quantifiers use this method, they should periodically check the proportion of dry solids in the slurry, measuring a half-dozen small samples every few months until the degree of variability is clear.

Quantifiers can use ratios observed from nonproject locations to estimate project-specific emissions per unit of manure input. However, this approach is valid only if all aspects of manure generation and handling at those locations are similar to those of the project, including the digestibility of the feed, the carbon and nitrogen content of the feed, and the types and ages of the animals. Such emissions estimates usually calculate emissions as a function of dry mass of manure input, possibly considering other factors. As noted, even if quantifiers use a published ratio to predict emissions as a function of manure inputs and handling practices, they must often convert measurements of wet manure to dry mass, as emissions ratios are usually published in terms of dry-manure mass. In these cases, the project plan must include protocols for determining the project-specific dry-mass ratio.

Quantifiers using project-specific ratios of emissions to manure inputs can also rely on the IPCC's default estimates of the amount of manure produced by each animal to estimate total emissions. If quantifiers use more than one method to calculate emissions and the results substantially agree, their analysis has greater credibility.

Calculating Methane and Nitrous Oxide Emissions

Quantifiers should develop a project-specific rate of methane emissions per unit of manure input. They can then measure the amount of manure input during each accounting period and multiply that by the emissions rate to find the total emissions for that period. If manure input is measured by volume, based on either flow or changes in the amount in a container, the ratio converts this volume to mass of GHG emitted from the manure. If the input is measured in mass, the ratio predicts the mass of gas emitted.

For example, suppose a project includes 100 dairy cows. Sampling shows that an open lagoon emits 0.145 metric tons of methane per ton of manure dry matter. Further, suppose that measurements show that each dairy cow produces 1.9 metric tons of dry matter[9] every year, for a project total of 190 tons. Thus the open lagoon emits 27.55 metric tons of methane that year. These emissions have a warming potential of about 579 tons of CO_2e (based on the IPCC's [1995] 100-year global warming potential for methane of 21).

Quantifiers can compare these project-specific calculations to the IPCC's default annual emissions per animal, by type and region. The IPCC's default maximum emission for dairy cows in developed countries is 0.1608 metric tons of methane per metric ton of manure dry matter (see Table A.16 in Appendix 18). This default maximum rate is about 11 percent more than the hypothetical measured rate of 0.145 tons of methane per ton of dry matter input. If the project does not leave manure in the lagoon as long as is typical, or if the lagoon temperature is cooler than average, this modest difference is logical, and the default rate would support the conclusion that the measured rate is accurate. If instead the project places manure in the lagoon only during fall and winter, and it spreads the manure on fields in the spring and summer, quantifiers would have to calculate the emissions from each practice and sum them to find the total emissions.

As when calculating methane emissions, quantifiers determine the project-specific rate of nitrous oxide emissions per unit of manure input. They then multiply the measured amount of manure input by that rate. As noted, the global warming impact of these emissions is usually quite small compared with that of methane.

For example, consider again the project with 100 U.S. dairy cows that stores manure in an open lagoon for years. Suppose the ratio of nitrous oxide emissions is 0.000105 metric tons per ton of manure dry matter. The lagoon receives 190 metric tons of manure dry matter during the year. Therefore, 0.01995 metric tons of nitrous oxide are emitted that year. These emissions have a warming potential of about 6.2 tons of CO_2e (based

on the IPCC's [1995] 100-year global warming potential for nitrous oxide of 310). This is just over 1 percent of the warming effect of the methane emitted by the lagoon.

Rather than determining and using a project-specific ratio, quantifiers can estimate nitrous oxide emissions based on manure mass, nitrogen content, and management practices. University and other laboratories perform dry combustion analysis of samples for a modest fee. (See Appendix 18 for specific equations used to calculate methane and nitrous oxide emissions from emissions ratios and manure inputs.)

Accounting for Carbon Dioxide Emissions

Manure projects emit CO_2 when the manure decomposes and when combustion destroys methane. Quantifiers may count CO_2 emissions when calculating both project and baseline emissions, or they may exclude those emissions from both. Not counting CO_2 emissions from decomposing manure is acceptable because the carbon content comes from crops that remove CO_2 from the air. In fact, nearly all the carbon in CO_2 emitted from manure had been absorbed from the atmosphere by plants 1 to 4 years prior, often within the same year. Given this relatively fast cycling, counting carbon content in growing crops as sequestration is unreasonable. If the carbon in crops does not count as sequestration, there is no reason to count the release of that carbon as emissions.

When calculating the GHG benefits of a project that converts methane to CO_2, quantifiers need not account for the CO_2 produced in the conversion. Thus, destroying about 1 ton of methane would avoid a warming effect of 21 tons of CO_2 (based on the 1995 GWP of methane of 21).[10]

Calculating Fugitive and Other Emissions

Quantifiers should include fugitive emissions when calculating project and baseline emissions. Such emissions occur most often during storage of manure, when containment vessels are not totally airtight; during transfer of manure; and as a result of incomplete combustion.

Some types of projects may find that most emissions are fugitive emissions. For example, projects with anaerobic digesters that flare methane produced by the digester would have no other methane emissions if combustion were complete and there were no leaks in the system. However, flaring does not completely combust all methane fed into the flare, so all facilities have fugitive emissions. Handling and storage of waste before and after digestion may also produce emissions. The IPCC gives a default fugitive emissions rate from digesters of 5 to 15 percent.[11] However, quantifiers can use sampling to actually measure fugitive emissions. If they do not use sampling, and if the digester is engineered and inspection shows it to be properly constructed and operated, quantifiers can conservatively assume that fugitive emissions are 10 percent of the methane produced. A default estimate for an engineered flare is that 98 percent of the methane is combusted. A default combustion rate for farm-designed and farm-built flares is 95 percent. Generators typically combust at least 99 percent of the methane input.

A manure project can have a high likelihood of producing indirect GHG emissions. If a project uses more electricity to operate a new aerobic digester, for example, quantifiers must account for the emissions from that increment of electric power. Some digester designs also include spraying, which consumes substantial amounts of energy. However, if a project sells sludge as fertilizer, that would probably not increase emissions, as the purchaser would probably have acquired fertilizer from another source if the project did not sell it. Such use would actually decrease emissions if the sludge had fewer emissions from manufacturing and use, per unit of nutrient, than the alternative form of fertilizer. However, the project could probably not take credit for those reduced emissions because it would not own them.

Determining Baseline Emissions

The method required to determine baseline emissions for manure-handling projects can differ considerably from that required for other types of projects. Quantifiers may establish baseline emissions of methane and nitrous oxide on comparison lands when writing the monitoring and verification plan, or they may simply describe how they will determine the baseline.

Calculations of baseline emissions include three components:

– Determining the manure-management practices that probably would have been used in the absence of the project.
– Determining the rate of emissions from each practice.
– Multiplying the rates by the amount of manure generated by the project.

Quantifiers can use EPA data to calculate average emissions for the dry weight of manure, weighted by the prevalence of each management practice in the relevant state. They can multiply this rate by the project's manure production to calculate baseline emissions for each accounting period. If local data are not available, quantifiers can use IPCC default factors to estimate emissions rates for each practice.[12]

Rates of methane production from lagoons (as a proportion of carbon input) vary widely, depending on how long the manure stays in the lagoon and the temperature of the material. Quantifiers should choose a conservative rate that is appropriate to the project to allow for uncertainty.

Changes in manure-handling practices on lands similar to the project should serve as the baseline. These shifts might include changes in the age or feed of the animal population, as well as changes in manure-handling equipment. Quantifiers can consult information from the EPA to determine the rates of use of different manure-management practices on comparison lands. Rather than conducting detailed on-site measurements of emissions from these comparison facilities, quantifiers can rely on mean estimates from agencies such as the EPA.[13]

To determine a hypothetical baseline, consider an anaerobic digester project with a constant rate of manure input. Before the project, the land managers in the region stored manure in open lagoons, which emitted 40 tons of methane per year. In the tenth year of the project, 10 percent of comparison facilities formerly using open lagoons had switched to digesters, 10 percent

Table 8.2 Weighted Average of Baseline Emissions in Year 10 of a Hypothetical Project, Compared with Project Emissions

Management practice	Emissions	Probability of practice	Weighted emission
Anaerobic digester	5	0.10	0.5
Daily spreading	1	0.10	0.1
Lagoon, covered and flaring	10	0.40	4.0
Lagoon, open	40	0.40	16.0
Baseline total		1.00	20.6
Project: Anaerobic digester	5	1.00	5
Project: Direct emission reduction			15.6

Note: Emissions are in tons of methane per year.

had switched to daily spreading, 40 percent had covered the open lagoons and flared the methane, and 40 percent were still using the open lagoons.

Assume that all the facilities are the same size and process the same volume of manure. Assume also that the digester emits 5 tons of methane annually, daily spreading emits 1 ton, and the system for covering and flaring emits 10 tons. The quantifier would find the weighted average by multiplying, for each possible outcome, the probability of that outcome by the emissions for that outcome and then summing. This produces a baseline emission of 15.6 tons of methane (see Table 8.2). Conversely, if all other facilities using open lagoons continued to use them, the baseline emission would remain constant at 40 tons of methane per year. In this case, the net emissions reduction for a project that only used digesters would be 40 − 5 = 35 tons annually.

The baseline of manure projects is often misunderstood. Suppose a project switches from an open pit system to a digester and flares the methane produced by the digester, while none of the animal operations in the region change their manure-handling systems. Further suppose that the pit system produces 5 tons of methane per month, and emits all the methane to the atmosphere. The digester, with the same manure input, produces 10 tons of methane per month and burns 9 of those tons. The system leaks the remaining ton as

fugitive emissions or with exhaust because of incomplete combustion. A common error is to assume that the 10 tons of emissions from the digester are the baseline and that the project benefit is the 9 tons per month that are burned. However, the baseline is the emissions from the pit system, or 5 tons per month. The emissions benefit is the baseline methane emissions (5 tons) minus the project methane emissions (1 ton), yielding a net benefit of 4 tons of methane per month.

Because baselines for manure projects often change over time, quantifiers need to establish them for each accounting period. If a project establishes baseline emissions before it starts, the monitoring plan should specify conditions under which quantifiers must recalculate them to account for unexpected changes, and should ensure that data are available to support recalculation.

Quantifying Leakage and the Bottom Line

Manure-management projects usually do not cause leakage, as they do not reduce the amount of animal products produced from project lands. However, projects that increase the export of manure could displace emissions to lands outside the project boundary. If projects increase the amount of manure a farm exports, quantifiers should account for emissions from those additional exports.

If projects spread sludge on fields outside the project boundary, quantifiers must also estimate emissions from the sludge that is spread and count them as leakage. Unless the sludge is saturated, methane emissions should be negligible, but nitrous oxide emissions could be significant. The default assumption is that 2 percent of the nitrogen in sludge is converted to nitrous oxide and emitted.[14] Quantifiers can use molecular weights to determine that 1 ton of nitrogen converted to nitrous oxide generates $44/28$ = about 1.57 tons of nitrous oxide.

If a project acquires manure by reducing the amount applied to nonproject lands, and if the amount of carbon in soil on those nonproject lands falls as a result, the project should account for that loss. However, off-site sequestration may not fall if the imported manure was formerly stored in a lagoon, composted, burned, or landfilled. Off-site sequestration may also remain steady if the amount formerly applied to lands greatly exceeded the amount that plants and microbes could process.

To calculate offsets, quantifiers subtract on-site emissions—including fugitive emissions and any downstream emissions, such as nitrous oxide from nitrogen in discharged effluent—from the baseline, adjusting this amount for any leakage.

Projects based on changing manure-handling systems to reduce the emissions of methane are relatively straightforward. Besides lowering GHG emissions, many of these projects allow landowners to reduce air and water pollution that can stem from decomposing manure. Moreover, the use of captured methane to power an electric generator or similar device offers additional opportunities for the landowner to reduce dependence on fossil fuels and reduce costs. Projects based at facilities that generate large amounts of manure, such as confined-animal feeding operations, can prove especially lucrative as well as environmentally beneficial.

Chapter 9

Step 6: Quantifying and Minimizing Methane and Nitrous Oxide Emissions from Soil

Soils and decaying organic material (including litterfall in forests, mulch used on croplands, and manure stored in lagoons) can emit both methane and nitrous oxide. Because the GWP of these two trace gases are high—23 and 296, respectively—these emissions increase potential complications as well as opportunities for landowners and developers of offset projects. Some activities that aim to sequester carbon in soil can lead to inadvertent emissions of methane and nitrous oxide, reducing the amount of offsets a project creates (see Appendix 2). On the other hand, landowners and project developers can adopt practices that reduce these emissions and thus increase the amount of offsets credited to their project.

Many factors affect methane and nitrous oxide emissions from soil (see Appendix 19), and the impact of those factors can vary widely from one point to another a couple of centimeters away. Efforts to measure changes in such emissions from fields and more extensive lands are therefore difficult and costly (see the sidebar). Reliably measuring annual emissions of these gases from even one field can cost $50,000 to $150,000—far more than the value of the offsets most projects can generate. Fortunately, two approaches that do not require field measurements are available for estimating changes in methane and nitrous oxide emissions from soil.

Using Simple Equations to Estimate Emissions

One approach to estimating methane and nitrous oxide emissions relies on the *denitrification-decomposition process model*, or DNDC (Li, Frolking, and Frolking 1992; Li, Narayanan, and Harriss 1996; Li, Aber, Stange, Butterbach-Bahl, and Papen 2000; and Li 2001). The GHG Wizard version of DNDC uses data provided with the model on the weather, soil types, and crop types/acreage of each county in the United States, as well as user-specified data on fertilization, tillage, and other management practices for each crop rotation and year. The model uses this information to estimate changes in soil carbon, changes in methane and nitrous oxide emissions, and the global warming equivalents of these emissions. (See http://www.dndc.sr.unh.edu/ for the model, instructions on its use, and detailed discussions of its applications.) However, this model requires a great deal of site-specific information—perhaps more than a project with diverse fields can easily provide—as well as expert judgment. Moreover, the practices and inputs specified by the model may differ from those of the project.

The second approach is to use simple equations to estimate emissions based on a few factors that are relatively easy to measure, such as the amount of carbon in soil, the amount of nitrogen land managers apply, and the demand for nitrogen by crops. Tables 9.1 and 9.2 provide examples of such equations, which are based

Measuring Methane and Nitrous Oxide Emissions in the Field

Field measurements provide direct observations of the rates, or fluxes, of GHG emissions from soils. Two general approaches are available to measure such fluxes: chamber sampling and open-air gas sampling.

Chamber sampling systems are open-bottomed containers that are sealed to the soil and closed for a limited period, with the rate of change in the concentration of gas in the chambers measured over time. The chambers must be of a known volume and cover a known area of soil so quantifiers can calculate the amount of gas emitted from the soil per unit of area and time.

Open-air gas-sampling systems take gas samples from points above the ground and use complex diffusion equations to calculate the emissions rates that would have caused the observed concentrations of gas. Open-air sampling can be used to estimate emissions over large areas, but it can be confounded by weather and other uncontrollable factors. Chamber systems, in contrast, are not unduly influenced by uncontrollable factors, but they sample only a single patch of land at a time. Because methane and nitrous oxide emissions can vary greatly from one locale to another, this can be a major drawback and usually dictates the use of many chambers to sample an entire project area.

Chamber systems have been widely used to take field measurements of methane and nitrous oxide emissions rates at local scales, from about 0.1 square meters to several square meters. Chambers are easy to make and cost little, but they are usually labor intensive to operate. Measurements must be extremely accurate because the changes in gas concentrations are very small. Moreover, if the gas concentration in the chamber increases much above the ambient level, diffusion of gas from the soil will slow.

Chambers are usually closed for periods lasting from a few minutes to an hour or two. The chambers are left open between sampling times to limit their impact on soil temperature, humidity, and plant growth. Each chamber is usually sampled four or more times per day during times of high emissions, such as after a major rainfall. Sampling may be infrequent during periods of low emissions, such as in the middle of a dry season or when the ground is frozen.

High-frequency sampling using automated chambers provides the best data for quantifying seasonal and annual emissions rates. Such chambers can collect samples every few hours and hold them, allowing crews to transport the stored samples to a laboratory for analysis every few days, limiting labor costs. Automated chambers are especially valuable for measuring nitrous oxide because most such emissions occur during a few pulses each year lasting one to five days, and sampling weekly or less frequently can miss those pulses. However, automated chamber systems can cost as much as $125,000.

Several technologies are available for quantifying the concentrations of methane and nitrous oxide in gas samples. These include gas chromatography with electron capture detection, photoacoustic infrared spectrometry, thermal conductivity detection, and tunable diode laser spectroscopy. Methane is often measured using gas chromatography with a flame-ionization detector. However, such analyzers cost tens of thousands of dollars and require trained technicians to operate. Still, these technologies are improving, and easier-to-use technologies may emerge. As an alternative, many university and commercial laboratories will analyze gas samples for a few dollars each. TRAGNET, a group of organizations that measure emissions rates of trace gases from soil, provides information to qualified users (see http://www.nrel.colostate.edu/projects/tragnet/).

Open-air gas-sampling techniques average emissions across areas up to tens of kilometers, depending on air turbulence and the height at which samples are taken. Converting observed concentrations of gas to emissions requires calculating diffusion rates, which depend on the texture of the ground surface, turbulence, temperature differences between the ground and atmosphere and within the atmosphere, and other factors. See Galle et al. (2000) for a review of the various techniques that can be used to calculate emissions rates from open-air sampling.

Table 9.1 Equations Derived from Results of DNDC Simulations to Estimate
Methane Emissions from Rice Paddy Soils in the United States

Parametric equation	$$F = \prod_{i=0}^{7} A_i \times clay^{B_0}$$
Coefficient equations	$A_0 = 108.61$ $A_1 = (F{-}2D + 2.4711)/80.464$ $A_2 = 0.0106T^{1.5021}$ $A_3 = 1.5127\text{Ln}(PH){-}1.8492$ $A_4 = 0.185\text{Ln}(Y){-}0.6204$ $A_5 = 7\text{E}{-}05(MD)^{-0.2605}(MA) + 0.9989$ $A_6 = 19.612(SOC)^2{-}0.2877(SOC) + 0.9834$ $A_7 = {-}0.5669\text{Ln}(LEAK) + 1.8394$ $B_0 = {-}0.73$
Definitions	F: soil CH_4 flux in rice growing season, kg CH_4–C/ha $clay$: soil clay fraction A_{0-6}: coefficients, kg CH_4–C/ha B_0: coefficient F: flooded days during the rice growing season D: drained days during the rice growing season T: mean annual air temperature, ºC PH: soil pH MD: days of manure amendment before start of flooding MA: amount of manure amended, kg manure–C/ha SOC: soil organic carbon content, kg C/kg soil $LEAK$: soil water leaking rate, mm/day Y: crop yield, kg C/ha/growing season

Notes: The symbol $\prod$ is used to denote the product of a series in much the same way that $\sum$ is used to denote the sum of a series. Thus, $\prod_{i=0}^{3} X_i = X_0 \times X_1 \times X_2 \times X_3$.

on results from the DNDC model that are fitted to different inputs.

Quantifiers can use the equations in Table 9.1 to estimate changes in *methane* emissions from rice paddies resulting from changes in

- The duration of flooding and drainage during the growing season.
- The amount of time between manure application and flooding.
- The amount of carbon added to the soil as manure.
- The carbon content of the soil.
- The acidity (pH) of the soil.

Quantifiers can use the equations in Table 9.2 to estimate changes in *nitrous oxide* emissions resulting from changes in

- The application rate of the nitrogen fertilizer.
- The application rate of carbon in the manure.
- The amount of organic carbon in the topsoil.
- The crop demand for nitrogen.
- The water input from precipitation and irrigation.
- The average annual air temperature.
- The clay content of the soil.
- The acidity (pH) of the soil.
- The land use (cropland, rice paddy, or grassland).

Table 9.2 Equations Derived from Results of DNDC Simulations to Estimate Nitrous Oxide Emissions from Soils in the United States

Total annual soil N_2O flux, kg N_2O-N/ha/yr	$$F = \prod_{i=0}^{3} A_i + \prod_{j=1}^{2} B_j \frac{R_f}{\prod_{k=0}^{6} K_k + R_f}$$
Coefficient equations	$A_0 = 1/(LU)$ $A_1 = 245C - 1.4385$ $A_2 = 1E{-}05(CN)^2 - 0.0053(CN) + 1.5254$ $A_3 = 0.9259e^{0.0005(M)}$ $B_1 = 0.2207e^{0.1858(T)}$ $B_2 = 21.704{*}\ln C + 122.51$ $K_0 = 300$ $K_1 = 0.2356e^{0.1694(T)}$ $K_2 = 1E{-}05(P)^2 - 0.004\,(P) + 5.5656$ $K_3 = 1.0339e^{3.9509(CLAY)}$ $K_4 = 0.2029(PH)^2 - 2.7911(PH) + 10.568$ $K_5 = 0.0745e^{0.0166(CN)}$ $K_6 = -9E{-}05(M) + 0.9808$
Definitions	F: annual soil N_2O flux, kg N/ha/yr R_f: fertilizer application rate, kg N/ha/yr A_{0-3}: background N_2O flux coefficients, kg N/ha/yr B_{1-2}: saturated N_2O flux coefficients, kg N/ha/yr K_{0-7}: rate coefficients C: SOC content in top soil, kg C/kg soil CN: crop demand for N, kg N/ha M: manure application rate, kg C/ha T: mean annual air temperature, °C P: total annual precipitation, mm $CLAY$: soil clay fraction PH: soil pH LU: land-use type (cropland 1, rice paddy 2, grassland 3)

* *Notes*: The symbol $\prod$ is used to denote the product of a series in much the same way that $\sum$ is used to denote the sum of a series. Thus, $\prod_{i=0}^{3} X_i = X_0 \times X_1 \times X_2 \times X_3$.

Opportunities to Minimize Methane and Nitrous Oxide Emissions

While some project activities can lead to inadvertent emissions, project managers can also adopt land-management practices that *reduce* emissions of methane and nitrous oxide. For example, converting degraded agricultural systems to perennial grasslands or forests can not only add substantial amounts of carbon to the soil, but also cut nitrous oxide emissions and enable the soil to absorb more methane. Reductions in irrigation may further reduce methane and nitrous oxide emissions from the soil (see Tables 9.3 and 9.4). Just as quantifiers must take emissions increases in account, they can take credit for these emission reductions when determining the net amount of offsets a project produces.

Table 9.3 Options for Mitigating Methane Emissions from Soil

Target factor to change	Mitigation option	Feasibility
Oxidizing capacity of soil	*Action*: Decrease number and duration of floodings of rice field soils. *Effect*: Periodically elevate soil redox potential. *Result*: Reduce CH_4 emissions, save water, increase crop yields, but increase N_2O emissions.	**
	Action: Apply oxidants (e.g., nitrate, Mn^{4+}, Fe^{3+}, sulfate) to wetland soils. *Effect*: Elevate soil redox potential temporarily. *Result*: Reduce CH_4 emissions.	*
	Action: Increase soil aeration by converting wetland to upland. *Effect*: Elevate soil redox potential permanently. *Result*: Reduce CH_4 emissions, increase N_2O emissions, often reduce carbon sequestration.	*
	Action: Loosen compacted soils in grazed pastures. *Effect*: Elevate soil aeration during rainfall events. *Result*: Reduce CH_4 emissions, increase CO_2 emissions.	*
Dissolved organic carbon (DOC)	*Action*: Reduce organic matter (litter or manure) incorporation in wetland soils. *Effect*: Decrease DOC with low decomposition rates. *Result*: Reduce CH_4 emissions, reduce carbon sequestration.	*
	Action: Incorporate organic matter in low-quality in wetland soils. *Effect*: Decrease DOC by reducing decomposition rates. *Result*: Reduce CH_4 emissions, but may reduce soil fertility.	*
	Action: Develop new rice cultivars with low root mass or exudation rates. *Effect*: Decrease root-produced DOC. *Result*: Reduce CH_4 production.	*
	Action: Reduce plant biomass in natural wetlands. *Effect*: Decrease root-induced DOC. *Result*: Reduce CH_4 production.	*
	Action: Apply crop straw or manure in rice paddies before transplanting or after harvest. *Effect*: Decrease availability of straw-decomposition-induced DOC to methanogens. *Result*: Reduce CH_4 production.	**
Gas transport	*Action*: Replace vascular plants with nonvascular plants in natural wetland. *Effect*: Eliminate pathway for gas transport. *Result*: Reduce CH_4 emissions.	*
	Action: Replace rice breeds with cultivars that have barriers to gas transport in stems and roots. *Effect*: Reduce pathway for gas transport. *Result*: Reduce CH_4 emissions.	***

Notes: * Substantial negative environmental effects in most locations; often too expensive.

** Negative environmental effects in some situations; often too expensive.

*** Typically no negative environmental effects; financially feasible.

See Appendix 19 for information on the oxidizing capacity of soil.

Table 9.4 Options for Mitigating Nitrous Oxide Emissions from Soil

Target factor to change	Mitigation option	Feasibility
Oxidizing capacity of soil	*Action*: Loosen compacted soils in grazed pastures. *Effect*: Elevate soil redox potential during rainfall or irrigation events. *Result*: Reduce denitrification-induced N_2O, but affect C sequestration.	*
	Action: Change soil texture by adding sand or silt or clay. *Effect*: Alter soil aeration status. *Result*: Reduce denitrification-induced N_2O, but could affect C sequestration.	*
	Action: Convert cultivated organic soils into wetland conditions. *Effect*: Build up deeply anaerobic conditions. *Result*: Eliminate both nitrification- and denitrification-induced N_2O, but could increase CH_4 emissions.	*
	Action: Reduce frequency of flooding and drainage cycles in wetland soils. *Effect*: Reduce soil decomposition and nitrification, but enhance denitrification. *Result*: Eliminate both nitrification- and denitrification-induced N_2O, but could increase CH_4 emissions.	*
Dissolved organic carbon (DOC)	*Action*: Reduce organic matter (litter or manure) incorporation in soils. *Effect*: Decrease DOC with low decomposition rates. *Result*: Reduce both nitrification- and denitrification-induced N_2O, but affect carbon sequestration.	*
	Action: Decrease quality of organic matter inputs. *Effect*: Decrease DOC by reducing decomposition rates. *Result*: Reduce both nitrification- and denitrification-induced N_2O, and maintain carbon sequestration.	*
	Action: Convert the upland with organic soils into wetland. *Effect*: Decrease DOC by depressing decomposition. *Result*: Reduce both nitrification- and denitrification-induced N_2O, may reduce CO_2 emissions from loss of soil organic carbon, but may increase CH_4 emissions.	**
Nitrogen	*Action*: Optimize nitrogen fertilizer application rates based on soil fertility and crop demand. *Effect*: Reduce nitrogen availability to nitrifiers and denitrifiers. *Result*: Reduce nitrification- and denitrification-induced N_2O, increase fertilizer efficiency, maintain optimum yield.	***
	Action: Apply nitrification inhibitors. *Effect*: Reduce nitrate availability to denitrifiers. *Result*: Reduce nitrification- and denitrification-induced N_2O.	**
	Action: Precisely schedule the timing of fertilizer applications. *Effect*: Increase fertilizer efficiency, reduce nitrogen availability for nitrifiers and denitrifiers. *Result*: Reduce nitrification- and denitrification-induced N_2O.	**
	Action: Rotate crops to reduce excess nitrogen in soils. *Effect*: Increase fertilizer use efficiency, reduce nitrogen availability for nitrifiers and denitrifiers. *Result*: Reduce nitrification- and denitrification-induced N_2O.	**

Table 9.4 Options for Mitigating Nitrous Oxide Emissions from Soil (continued)

Target factor to change	Mitigation option	Feasibility
	Action: Use control-release fertilizers. *Effect*: Increase fertilizer use efficiency, reduce nitrogen availability for nitrifiers and denitrifiers *Result*: Reduce nitrification- and denitrification-induced N_2O.	**
	Action: Convert upland with organic soils to wetland. *Effect*: Depress nitrification and accelerate denitrification. *Result*: Reduce both nitrification- and denitrification-induced N_2O, but increase CH_4 emission.	*
	Action: Convert intensive to reduced tillage. *Effect*: Reduce nitrogen availability by decreasing mineralization rates. *Result*: Reduce near-term nitrification- and denitrification-induced N_2O, but may increase long-term N_2O emissions due to elevated soil organic carbon.	*
	Action: Compost organic materials before applying to soil. *Effect*: Reduce mineralization and consume free nitrogen in soil. *Result*: Reduce both nitrification- and denitrification-induced N_2O. *Note*: Use anaerobic digesters to pretreat manure or crop residue to eliminate labile nitrogen.	**
Temperature	*Action*: Plant winter cover crops or use mulch in winter. *Effect*: Moderate surface soil temperature. *Result*: Reduce freezing/thawing-induced N_2O.	**
pH	*Action*: Convert cropland or grassland to selected conifer tree species. *Effect*: Reduce soil pH to 5.0 or lower. *Result*: Reduce both nitrification- and denitrification-induced N_2O.	**

Notes: * Substantial negative environmental effects in most locations; often too expensive.
** Negative environmental effects in some situations; often too expensive.
*** Typically no negative environmental effects; financially feasible.
See Appendix 19 for information on the oxidizing capacity of soil.

Although methane and nitrous oxide are often not considered in the context of global warming and the "low-carbon" economy, they are potent GHGs. These emissions from soil provide landowners and project developers with both a challenge and an opportunity. Some project activities, such as no-till farming, can increase emissions of these gases and reduce an offset project's bottom line. However, other practices can reduce these emissions and increase the bottom line. In either case, quantifiers must account for and estimate changes in methane and nitrous oxide emissions when assessing a project's net greenhouse impact.

Chapter 10

Step 7: Estimating Leakage or Off-Site Emissions Caused by a Project

Leakage refers to changes in GHG emissions or carbon stocks that occur outside a project's boundary but that nevertheless can be attributed to the project's activities. Projects whose emissions are directly capped by a regulatory system do not need to account for leakage.[1] However, projects designed to produce marketable offsets must subtract these emissions from their net GHG benefit.

Leakage usually occurs when a project reduces the supply of a good, displacing production—and thus GHG emissions—to another location.[2] Two types of projects might displace production. The first type reduces production to reduce GHG emissions. Examples include ending logging to avoid CO_2 emissions from harvested timber and ending rice cultivation to stop methane emissions from flooded rice paddies. The second type occurs when a project establishes a new land use, displacing some preexisting use. An example is a forestation project on formerly agricultural land. In both types of projects, reductions in the supply of forest or agricultural products can boost production elsewhere, reducing the overall benefit of the projects.[3, 4] This chapter outlines the methods projects can use to quantify these types of leakage. (See Appendices 20–23 for more information.)

Leakage as a Function of Supply and Demand

Although project developers may not know exactly where leakage occurs, they can calculate it by consider-

ing the relative sensitivities of the quantities of goods supplied to a market—and the quantities of goods demanded by buyers—when the price of those goods changes. In the language of economists, this involves determining to what extent supply and demand are *elastic* or *inelastic*.

For nearly every terrestrial offset project, leakage results from a change in the supply of a product or good, not from a change in demand.[5] This change can be a decrease in supply, such as when a forest-conservation project removes land from the timber base, or an increase in supply, such as when an afforestation project includes a harvest that increases the amount of timber on the market.

When a project reduces emissions by reducing the supply of a good, other suppliers may compensate for a portion of the lost production and hence replace a portion of the reduced emissions. The proportion of the cut that others compensate for depends on the relative sensitivity of suppliers and consumers to changes in the price of the good. The rate of change in the amount of the good supplied as a function of a change in price is called the *price elasticity of supply*. The rate of change in the amount of the good demanded by consumers as a function of a change in price is called the *price elasticity of demand*.

Formally, the price elasticity of supply is the fractional change in supply for a given fractional change in price:

$$e = \frac{\left(\dfrac{\Delta Q_S}{Q_S}\right)}{\left(\dfrac{\Delta P}{P}\right)}$$ Equation 10.1

where e is the price elasticity of supply, Q_S is the quantity of the product supplied before the price change, ΔQ_S is the change in the quantity of the product supplied induced by a price change, P is the price of the product received by suppliers before the price change, and ΔP is the change in the price that suppliers receive for the product. $\Delta Q_S/Q_S$ and $\Delta P/P$ are the fractional changes in supply and price, respectively. Generally, supply increases as price increases. Thus the price elasticity of supply is usually a positive number.

Similarly, the price elasticity of demand is the fractional change in demand for a given fractional change in price:

$$E = \frac{\left(\dfrac{\Delta Q_D}{Q_D}\right)}{\left(\dfrac{\Delta P}{P}\right)}$$ Equation 10.2

where E is the price elasticity of demand, Q_D is the quantity of the product demanded before the price change, ΔQ_D is the change in the quantity of the product demanded induced by a price change, P is the price of the product paid by buyers before the price change, and ΔP is a change in the price buyers pay for the product. Demand for a product usually falls as the price

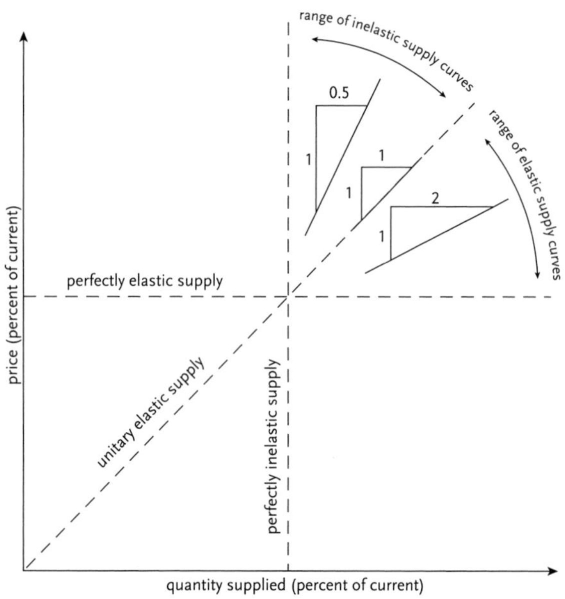

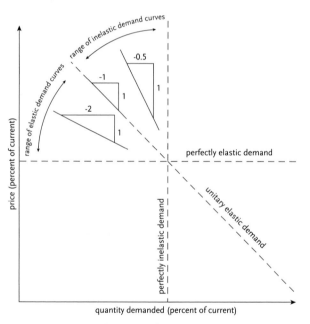

Figure 10.1 Price elasticity of supply is determined by plotting supply on the *x*-axis and price on the *y*-axis. The price elasticity of supply, *e*, is then the inverse of the slope of the line. The line "unitary elastic supply" shown here represents an elasticity of 1.0. That is, a 1% increase in price spurs a 1% increase in the amount of product supplied to the market. By convention, that is the dividing line between inelastic and elastic supply. The horizontal line represents a perfectly elastic supply ($e = \infty$). That is, any price shift induces an infinitely large change in the supply. The vertical line represents a "perfectly inelastic" supply ($e = 0$). That is, a shift in price does not induce any change in supply.

Figure 10.2 Price elasticity of demand is determined by plotting demand along the *x*-axis and price on the *y*-axis. The price elasticity of demand, *E*, is the inverse of the slope of the line. As in Figure 10, the lines represent unitary elastic demand, perfectly elastic demand ($E = \infty$), and perfectly inelastic demand ($E = 0$).

rises. Thus the price elasticity of demand is usually a negative number.

A graphical representation illustrates the derivation and significance of the price elasticities of supply and demand (Figures 10.1 and 10.2). Note that, in each case, the inverse of the slope of the line plotted in the graph equals the price elasticity (the inverse of the slope is $\Delta x/\Delta y$, where x represents the horizontal axis and y represents the vertical axis).

Economists classify supply and demand as elastic or inelastic depending on the values of e (for supply) and E (for demand). As noted in Figure 10.1, values of e between 0 and 1 are said to be inelastic, and values of e greater than 1 are said to be elastic. If there is no change in supply when a price changes, $e = 0$, and the supply is perfectly inelastic. If there is only one price for a good regardless of the supply, $e = \infty$, and the supply is perfectly elastic.

Similarly, as Figure 10.2 shows, values of E between 0 and –1 are classified as inelastic, and values of E lower than –1 are classified as elastic.[6] When there is no change in demand when a price changes, $E = 0$, and demand is perfectly inelastic. When there is only one price for a good regardless of the demand, $E = \infty$, and demand is perfectly elastic.

Calculating Leakage from Decreased Production Using Supply and Demand Elasticities

For projects that decrease production of some good, Murray, McCarl, and Lee (2004) calculate leakage as[7]

$$L = \frac{e \times C_{out}}{\left(e - \left(E \times (1 + \phi)\right)\right) C_{proj}}$$ Equation 10.3

where L is the proportion of the project's GHG benefit that is lost to leakage[8], e is the price elasticity of supply for the good provided by off-project producers, E is the price elasticity of demand for the final commodity, C_{out} is the amount of GHG emissions per unit of increased commodity production outside the project area, C_{proj} is the amount of GHG emissions eliminated per unit of reduced commodity production within the project area, and ϕ is the ratio of the market share of the project to the share of the rest of the market's producers.

Because the last parameter quantifies the effect of

the project on the market, it affects the amount of leakage. For most projects, market share (ϕ) is quite small and can be dropped from Equation 10.3. For example, if $\phi = 0.1$ (a relatively large number) dropping the term changes the leakage estimate by less than 2 percent, which is probably less than the error in elasticity estimates.

However, if the project supplies more than a small fraction of total market supply, it could have a discernable impact on the market's dynamics, and it should be retained in Equation 10.3. In that case, market share is calculated as

$$\phi = \frac{Q_{S,proj}}{Q_S}$$ Equation 10.4

where $Q_{S,proj}$ is the supply from the project area before the project began, and Q_S is the supply from nonproject lands (that is, total market production minus the amount provided from the project area).

Example: Reduced Timber Harvesting in the Pacific Northwest

To illustrate the use of Murray's formula, consider the impact on carbon sequestration of reducing timber harvest on federal lands in the U.S. Pacific Northwest to protect the spotted owl. These cuts began in the 1980s but occurred mostly in the early 1990s. Building on the work of Wear and Murray (2004), Murray et al. found a weighted-average U.S. supply elasticity of $e = 0.46$, a demand elasticity of $E = -0.06$, and a proportion of timber eliminated from the U.S. market by restrictions on logging on federal lands of $\phi = 0.045$. Wear and Murray did not analyze GHG emissions per unit of harvest, and Murray et al. simply assumed that $C_{out} = C_{proj}$.

Inspection of Equation 10.3 reveals that under this assumption, both C_{out} and C_{proj} drop from the equation (because $C_{out}/C_{proj} = 1$). Substituting into Equation 10.3,

$$L = \frac{(0.46 \times 1)}{\left(0.46 - (-0.06 \times (1 + 0.045))\right) \times 1} = \frac{0.46}{0.5227} = 0.88$$

This means that for every ton of emissions avoided by reducing timber harvest on federal lands, 0.88 tons of emissions occurred from increases in harvesting elsewhere. (The numbers used in this calculation are rounded; Murray et al. calculated leakage of 0.87.) Thus, if the reduced lumber harvesting program in the Pacific Northwest had been filed as a GHG offset project, the project could have claimed only 12 percent of its net greenhouse benefit as offsets (the benefit calculated after accounting for proportional additionality and the baseline). In other words, substituting into Equation 2.2,

$$\text{Offset} = \text{Net GHG Benefit} \times (1-0.88) = \text{Net GHG Benefit} \times 0.12 \qquad \text{Equation 10.5}$$

The relatively large amount of leakage in this example is due largely to the inelasticity of demand (E is only −0.06). This inelasticity most likely reflects the fact that even significant changes in log prices have only a moderate effect on consumer spending on goods produced from wood because of the need for housing material and paper.

Rules of Thumb for Leakage

The above example—in which the supply elasticity was larger than the magnitude of the demand elasticity[9] and leakage was large—is indicative of a rule of thumb. If emissions per unit harvested are the same on and off a project, leakage will vary proportionally with the difference in the price elasticities of supply and demand. When supply is more elastic than demand, leakage will be greater than 50 percent; when the opposite applies, leakage will be less than 50 percent (see Table 10.1). In this regard, Tables 10.2 and 10.3 suggest that supply elasticities are often greater than the magnitude of demand elasticities for terrestrial offset projects, and thus leakage is often greater than 50 percent for them. This shows how critical it is for such projects to account for leakage.

Another rule of thumb, revealed by Equation 10.3, is how leakage is affected by relative changes in C_{proj} and C_{out}. Leakage will decrease as emissions per unit of production within the project rise relative to emissions per unit of production in locations where displacement increased production. In other words, efficiency by landowners and project developers can decrease leakage and enhance the amount of offsets a project produces.

Market share can also influence leakage. Smaller projects usually have higher leakage rates than larger projects. That is because a change in supply from a small project has an insignificant effect on price, and thus it does not change consumption (a small increase in supply can meet existing demand). On the other hand, large projects that remove a large fraction of existing supply have limited leakage because the lack of finite resources such as land limits increases in production elsewhere.

Building on this general framework for calculating leakage, the next section outlines methods for calculating the elasticity and productivity values needed to calculate leakage with Equation 10.3. (See Appendix 22 for special issues in assessing leakage in afforestation projects.)

Obtaining Elasticity Numbers

Analysts can obtain supply and demand elasticities from the literature or calculate them. A variety of elasticities have been published for forest products and agricultural commodities (see Tables 10.2, 10.3, and 10.4). When using an elasticity value from the literature, analysts must ensure that it applies to the product in question. This means that the elasticity must

- Be for the exact product in question or a group of products that includes the product.
- Encompass the geographic area where the project is located.
- Be measured over a time period that is as close to the project time period as possible.

Table 10.1 Leakage as a Function of the Relative Elasticities of Supply and Demand

Relative price elasticity of supply and demand	Leakage
Supply less elastic than demand	Less than 0.5
Supply and demand equally elastic	0.5
Supply more elastic than demand	More than 0.5

Note: Leakage is expressed as a fraction of a project's overall GHG benefit.

Table 10.2 Selected Price Elasticities of Supply of U.S. Forest Products

Product/region of North America	Price elasticity of supply
Softwood lumber/Pacific Northwest, west side	0.335
Softwood lumber/Pacific Northwest, east side	0.586
Softwood lumber/Pacific Southwest	0.794
Softwood lumber/Northern Rockies	0.866
Softwood lumber/Southern Rockies	0.395
Softwood lumber/North Central	0.848
Softwood lumber/Northeast	0.188
Softwood lumber/South Central	0.937
Softwood lumber/Southeast	0.963
Softwood lumber/British Columbia coast	0.935
Softwood lumber/Canadian interior provinces	0.447
Softwood lumber/Canadian eastern provinces	0.492
Softwood plywood/Pacific Northwest, west side	0.748
Softwood plywood/Pacific Northwest, east side	0.444
Softwood plywood/Pacific Southwest	1.120
Softwood plywood/Northern Rockies	0.600
Softwood plywood/South Central	0.395
Softwood plywood/Southeast	0.343
Oriented strand board/United States	0.512
Oriented strand board/Canada	0.433

Source: Adams and Haynes (1996).

Table 10.3 Selected U.S. Price Elasticities of Demand

Product	Price elasticity of demand
Softwood saw logs	−0.34 to −0.44
Hardwood saw logs	−0.19 to −0.22
Pulpwood	−0.40
Cotton, domestic demand	−0.22
Cotton, export demand	−1.20
Corn, domestic demand	−0.23
Corn, export demand	−0.33
Wheat, domestic demand	−0.07
Wheat, export demand	−0.35
Sorghum, domestic demand	−0.20
Sorghum, export demand	−0.80
Wool, domestic demand	−0.40
Wool, export demand	−0.80
Soybeans, export demand	−0.82

Sources: Elasticities for forest products are from McCarl n.d. and elasticities for agricultural products from McCarl et al. (1993), both cited in Adams et al. (2005).

Table 10.4 Selected U.S. Price Elasticities of Supply

Product	Price elasticity of supply
Corn	0.16
Soybean	0.25
Wheat	0.2
Oats	0.4
Cotton	0.4
Timber	0.46

Source: Adams et al. (2005).

Because many agricultural commodities are marketed globally, it is important to specify global regions or nations that serve as markets for products from a project area. Detailed studies of market demand are available for many agricultural products in many nations (Seale 2003).

In the absence of existing estimates of price elasticities for project commodities, analysts can rely on estimates for groups of products that are similar to project commodities. However, they should pay particular attention to the characteristics of the product group. Because proxy estimates may be unreliable, analysts should use this approach only during scoping, not in finalizing a leakage analysis for a project.

Another approach is to calculate elasticities specifically for project products using established methods and data. Reliable estimates require analysis of multiple data points. Analysts can use a standard spreadsheet program to perform a simple linear regression of price against quantity and other important variables with multiple data points. More sophisticated regressions make it possible to assess the influence of other factors, such as changes in wealth or population, on supply and demand. Use of these other factors increases the reliability of the elasticity estimates. (See Appendix 23.[10])

Caveats and Special Considerations When Using Elasticities

Analysts should use caution when extrapolating supply and demand elasticities over large changes in prices or over long time periods. This section discusses the reasons why they should do so and describes the particular circumstances where landowners, project developers, and analysts need to pay extra attention.

With large decreases in price, one good may become a substitute for another good, and demand may become increasingly elastic. Demand may also become saturated and thus inelastic with further price decreases. With large increases in price, other goods may become substitutes, and demand for the good may become very elastic. For many goods, demand may become inelastic at high prices because the remaining buyers are strongly dependent on the goods or are sufficiently wealthy that the price does not matter. However, eventually, at some

very high price, limits to wealth will reduce purchases, and demand will become more elastic.

Similarly, the elasticity of supply often varies over large price ranges. Very high prices may draw production resources from other goods, and elasticity may increase. High prices may indicate some natural limit on the extraction of a natural resource, reflecting the inability of suppliers to substantially increase production beyond a certain level. Suppliers usually have some fixed costs of production, so at some low price, supply becomes inelastic as they stop making the good.

Elasticity of supply and demand may also change over time, so analysts should use caution when applying supply and demand elasticities to time periods that are years away from the data on which they are based. In the short run, elasticities tend to be lower, as firms and consumers cannot readily adjust production or consumption. In the longer run, these factors are not as fixed, and firms and consumers do adjust. Leakage studies usually require longer-run elasticities, with actors outside the project area adjusting production to changes stemming from a project.

For manufactured goods, in the absence of a change in technology, the long-run supply curve tends to be much more elastic than the short-term curve. This is because, over several years, suppliers can build new manufacturing plants to meet higher demand or retire old facilities to reflect a sustained drop in demand. The supply of natural resources may not be as elastic as that of manufactured goods. Increases in demand can encourage the exploration and development of new production or extraction technologies. However, for many resources, limits on access to raw material and amount used will eventually limit production. On the other hand, studies of the supply of wood products in the United States from the 1970s through 2000 have shown that elasticities are relatively stable (Haynes 2007).

Figure 10.2 showed demand with a constant price elasticity as linear (a straight line). In reality, demand curves with a constant elasticity are nonlinear: at different prices, a one-dollar change in price results in different changes in demand. This is because price elasticities are expressed as ratios of fractional changes (see Equations 10.1 and 10.2), and a one-dollar change in price is a larger proportion of a small price and a

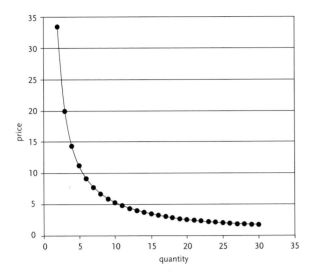

Figure 10.3 Demand as a function of price for a constant price elasticity of 1. While Figure 10.2 showed the demand with a constant price elasiticity as a straight line, demand curves with a constant elasticity are actually nonlinear as a function of price; this non-linearity becomes apparent when the plot is extended over a large price range. The nonlinearity arises because price elasticities are expressed as ratios of fractional changes (see Equations 10.1 and 10.2), and a one-dollar change in price is a larger proportion of a small price, and a smaller proportion of a large price.

smaller proportion of a large price. Figure 10.3 illustrates this effect. It plots demand and price for a constant elasticity of 1 over a large range of prices. Because of the nonlinearity, to estimate the effect on demand of a large change in price, analysts should construct a demand curve using logarithms of quantity and price. This will yield a straight line, making the extrapolation easier. However, analysts should use great caution in estimating the effects of large price changes because there is a substantial risk that elasticity will not remain constant.

Geographic scale also matters when calculating the price elasticity of supply. Whereas supply in a given region or small country may be constrained, many markets for goods are global. Another way of looking at this is that individual producers in large markets are price takers, and they have a limited ability to alter the prices they receive in the market. To a price-taking supplier, global demand appears to be almost infinitely elastic. If a market is very small because of a lack of substi-

tutes, legal or tax barriers to imports, high transportation costs, or other reasons, a single producer may face less-elastic demand than if the market were more open. In general, the more open and global the market, the smaller the project's market share will be and the larger the leakage will be.

Estimating C_{proj} and C_{out}

As already mentioned, C_{out} and C_{proj} are the emissions per unit of production (or unit of harvest) outside the project and within the project, respectively. Both appear in Equation 10.3 and thus are needed to calculate leakage.[11] It is especially important to keep in mind that, when using these parameters to calculate leakage, they must be in emissions per *unit of production,* not emissions per unit of area. Examples of emissions per unit of production include tons of carbon per board feet of wood and tons of carbon per bushel of grains.

In most cases, determining C_{proj} should be fairly straightforward. Because this parameter relates to production within the project area, the data needed to derive it should be readily available to the analyst.

Estimating C_{out} can be more problematic. Analysts can sometimes use studies of production costs and efficiencies, and some ecological and fire-fuel studies, to estimate emissions per unit of production outside the project area. In the absence of such data, analysts must estimate C_{out} from other sources. For example, they can use data from published tables, scientific papers, and models to estimate emissions per unit of area, as well as the number of units of the product made or harvested within a given area. C_{out} can then be obtained by dividing the latter into the former. Note that because units of area appear in the denominator in both terms, area cancels out, leaving emissions per unit of production.

When calculating leakage, the estimation accuracy of L in Equation 10.3 is enhanced by using the marginal rate of production[12] per unit of land area, rather than the average rate.[13] If data on amounts of production and area in production are available for multiple time periods, analysts can use that information to estimate the marginal rate of production. If they use two points in time, the equation for calculating the marginal rate of production is

$$R_{\text{marginal}} = \frac{(P_1 - P_0)}{(A_1 - A_0)} \qquad \text{Equation 10.6}$$

where R_{marginal} is the marginal rate of production of the good per unit area, P_1 is the amount of production at a later time, P_0 is the amount of production at an earlier time, A_1 is the area in production at a later time, and A_0 is the area in production at an earlier time.

The times for each set of P and A must be the same. Units of production (such as thousand board feet or bushels) must be the same for both Ps, and units of area (such as acres or hectares) must be the same for both As.

Analysts must check this estimate for reasonableness. Sampling and measurement errors, and year-to-year weather variability, can yield unreasonable estimates. If an estimate appears unreasonable, analysts should perform the calculation for a different time period. It may or may not be acceptable to average estimates made over several time periods, depending on the degree to which other economic factors are constant across those periods. As in the case of calculating elasticities, a more reliable (but more difficult) method is to use data from multiple times, include multiple variables, and use regression analysis to calculate the rate of production. Analysts can use a standard spreadsheet program to calculate a simple linear regression of quantity produced against area of land in production. (If more variables are addressed, analysts can consult Appendix 24 and other materials on this methodology.)

Leakage can often be a significant fraction of the net GHG benefit of a terrestrial offset project and thus can significantly reduce the amount of offsets that the project can achieve. For this reason, it deserves careful consideration by landowners, project developers, and analysts throughout the project. As a rule of thumb, leakage tends to decrease as

- Supply becomes more inelastic (that is, e becomes smaller).
- Demand becomes more elastic (E becomes larger in absolute value).
- Market share of the project increases (ϕ becomes larger).
- Emissions per unit of production from the project area become larger relative to emissions per unit of production in locations where displacement increases production (that is, C_{proj} becomes larger than C_{out}).

Landowners and project developers should be especially mindful of leakage when demand elasticity is less than supply elasticity, because under these conditions leakage is likely to be greater than 50 percent of the project's net GHG benefit. This is in fact likely to be the case for many terrestrial offset projects.

Chapter 11

Step 8: Verifying and Registering Offsets

When a qualified, independent party verifies a project's offsets, and the offsets are appropriately registered, buyers, sellers, regulators, and market observers receive assurance that the offsets represent real atmospheric benefits. Often, a verifier's essential function is to act as an auditor, attesting that a project's offsets have been quantified correctly (with suitable methods used to account for additionality, baselines, uncertainty, and leakage) and that ownership is as claimed. Subject to conflict-of-interest limitations, verifiers may also act as quantifiers, estimating the amount of offsets a project will generate as part of scoping the project, validating the project's monitoring and verification plan, and quantifying the amount of offsets a project produces. Of course, individual carbon-trading systems may require specific verification procedures or may require or prohibit specific relationships between verifiers and projects.

Projects should make information on the process of verifying offsets publicly available, most often through a registry—a central database of emissions and reductions. All regulatory GHG mitigation systems that rely on allowances or trading maintain an official registry, and many voluntary systems do as well. Each registry has an underlying accounting framework that

- Ensures that offsets meet a specified quality standard.
- Associates each offset with a unique owner.
- Removes expired or retired offsets from the market.

- Provides a central exchange where market participants can execute trades, if the system allows trading.

Assigning each offset in the registry a serial number or other identifier allows regulators and observers to trace its pedigree, even after numerous transfers from one party to another. This chapter addresses all these aspects of verification and registration. (See Appendices 24–26 for more detail.)

The Verifier as Auditor

If verifiers function as auditors, they verify the method, data, and calculations used to quantify offsets. In this role, verifiers may validate a project's monitoring and verification plan before it is used—attesting whether, if the project is applied, its results will show valid GHG offsets. Validation is recommended when project developers prepare the plan and collect and analyze project data. Early consultation is important to ensure that any measurements needed before monitoring and verification activities begin can be designed and conducted at that point.

After the validated quantification method has been applied, verifiers may then audit measurements and calculations performed by project developers, project managers, or other quantifiers such as a consulting engineering firm. (The Kyoto Protocol's Clean Development Mechanism generally prohibits the same party from both validating a quantification method and verifying offsets calculated using that method.) Some verifiers consider whether the calculated amount of offsets is reliable, and if it is not, they indicate a more reliable amount and explain why the verifier's calculation is more reliable. Other verifiers do not wish to take responsibility for the claimed amount of offsets, and they state only whether the quantifiers have followed the validated plan. (A carbon-trading system may not allow verifiers to charge a fee to attest only that the quantification plan has been followed without also expressing an opinion as to whether the claimed amount of offsets is real.)

Verification audits can range from light to intensive. If reports describing offset calculations are detailed, an audit could be as minimal as a review of the reports. However, such a light audit is not recommended, as it will not reveal whether data were collected correctly or whether measurements were properly converted for use in calculations.

A somewhat more intensive audit entails reviewing project documents, and it may include interviews with people involved in collecting data and making calculations. This mid-level audit allows verifiers to check the comprehensiveness of data and the accuracy of calculations.

An intensive audit entails visiting the project site to assess the appropriateness and reliability of data-collection methods and measurements, including whether instruments have been properly calibrated and used. Site visits may also help verifiers judge whether the extrapolation of measurements is reasonable. For example, if the solid content of manure slurry is measured at one facility and the proportion is applied to several facilities, a verifier should consider, "Do the facilities use similar enough methods for diluting manure to allow that extrapolation?"

Verifiers may also take measurements independently to check their accuracy. At a minimum, verifiers should repeat some calculations to see if they match reported results. Verifiers typically spot-check particular values, some randomly and others because they appear suspicious. The spot checks may yield slightly—and in some cases grossly—different values for reported emissions and sinks. If verifiers find that their values differ from the quantifier's values by less than a certain

threshold, and they do not suspect deliberate misrepresentation, they may be allowed to attest to the inventory *mutatis mutandis* ("the necessary changes being made"), that is, with minor revisions.

If not set by the registry or market regulator, the monitoring plan should explicitly state the threshold error. This defines the boundary between material and non-material misstatements. *Material misstatement* is a term of art in the field of accounting, defined by the Financial Accounting Standards Board as "the magnitude of an omission or misstatement of accounting information that, in the light of surrounding circumstances, makes it probable that the judgment of a reasonable person relying on the information would have been changed or influenced by the omission or misstatement" (AICPA 2004, AU312.10).

A more specific definition of material misstatement varies with the environment in which it is used. A market-wide threshold may be numerically complex to account for varying types of inventories and conditions, or it may rely on qualitative criteria. Any allowable threshold error should relate to the amount of offsets, not the size of emissions inventories. The maximum allowable error for offsets should be 5 percent of the quantified amount. If verifiers find discrepancies totaling more than that, the regulatory system should require revisions by the project developer and resubmission to the verifier. (See Appendix 24 for more on auditing functions and methods.[1])

A verification report should attest that a project's measurements and calculations are appropriate and correct to the level of precision in the quantification report. The verifier's report should include:

- The amounts of offsets found to have been achieved.
- At least a brief description of methods used by the quantifier.
- A description of how the verifier judged the appropriateness and accuracy of those methods.
- A description of any quantification errors or omissions.
- A description of how any quantification errors or emissions changed the amount of offsets.

–The verifier's opinion of how many tons of offsets the project achieved in each accounting period.

–A determination of whether emissions benefits qualify as offsets within the specified GHG mitigation system.

Auditing a quantification of GHG offsets requires substantial technical expertise and familiarity with the tools and methods used in the quantification. A typical financial accounting firm lacks the scientific, engineering, and economic expertise to perform such services. Moreover, a financial audit that merely determines that project records include numbers, and that arithmetic is correct, is not sufficient. The verifier must be able to judge whether the appropriate data were collected, whether the equations used in calculations are appropriate, and whether calculations of confidence levels are realistic.

The Verifier as Quantifier

In addition to functioning as auditors, verifiers may serve as quantifiers, determining what measurements projects should take, taking them, and calculating offsets. This approach is most appropriate when such efforts require significant technical capacity. When they do, project developers can reduce the cost of quantification by allowing individual verifiers to apply their own procedures and tools.

If verifiers function as quantifiers, they write the project's monitoring and verification plan (see the sidebar). A well-articulated plan specifies equations that will be used to calculate emissions and sinks. The plan should also specify the factors used in the equations or the methods for making site-specific measurements of each factor.

In theory, a quantifier should be able to apply a monitoring plan written by someone else and produce a reliable count of offsets. However, experienced quantifiers have instruments and methods with which they are familiar. A plan written by someone else may require them to acquire new measuring instruments, learn new methods, and build a new database or spreadsheet, thereby raising the cost. In addition, quantifiers can quickly calculate confidence intervals when applying

their standard methods to a new site, whereas it can require extensive work to determine the precision of new methods specified by someone else.

The process of valuing timber provides an analogy. Timber appraisers usually have years of experience meshing their timber-cruising methods with methods used to determine log grades and volumes according to local market standards. Timber appraisers also develop libraries of information and refine and calibrate equations to local situations. A data-gathering protocol designed by someone else would not provide the exact information an appraiser would need to perform the needed calculations. At best, the appraiser would have to spend extra time meshing the data with his or her methods, and the resulting calculation would probably not be as precise as if the appraiser had used a data-gathering design that matched the calculation methods. At worst, the cruiser would not be able to use the data to calculate timber value.

Establishing a Verification Agreement

A verification agreement specifies the obligations of the quantifier and the other contracting party. The agreement should

–Define the project or present a method for determining project boundaries.

–Specify a timeline for actions, such as taking measurements and delivering quantification reports.

–Provide a schedule of payments for the work.

–Determine procedures for dealing with unplanned occurrences.

The agreement may also specify the degree of precision to be achieved by quantification activities. If offsets are to be submitted to a particular GHG mitigation system, the agreement should specify that offset calculations shall comply with the requirements of the system. The agreement may require acceptance of the verifier's monitoring plan prior to implementation, and it may not run for the life of the project.

Before committing to quantifying a project, verifiers should do an initial check to ensure that the project can produce offsets. If the project cannot produce

Writing a Monitoring Plan and Ensuring Quality

Verifiers serving as quantifiers should create a monitoring and verification plan for each project that is concise but complete enough that others can comprehend it and accept offsets based on it. (A public version of the plan may contain less detailed information about management of project lands and their production.) The monitoring and verification plan should include:

- Project boundaries if they are known when the plan is written.
- A list of the project's potential sources or sinks of GHG emissions.
- A sampling design tailored to the project, including the locations of plot centers. The plan should specify three mechanisms for locating the plots, such as markings on aerial photos, GPS coordinates, and physical markers placed at the plots' reference points.
- Detailed protocols for measuring project conditions and outputs, including the frequency of measurements and procedures for recording, managing, and storing data.
- Procedures, factors, and equations for analyzing data and calculating project-specific ratios, and criteria for changing those procedures.
- The description of the contents and timing of reports that quantifiers will generate. These reports should include summaries of original data, not just the results of data analysis, so someone unfamiliar with the project can interpret the information and analysts later can apply new techniques as they become available.
- Quality-control standards and methods, including redundancy in recording data in case records are lost.
- A baseline or process for setting the baseline.
- Specification of a process for quantifying leakages.

Quality control typically entails requiring a second independent measurement of some plots, or reanalysis of some data, to see if the two measurements match. It also entails ensuring that quantifiers and verifiers consider multiple sources of information on project outcomes and that they investigate irregularities in project records.

If quantifiers plan to sample GHGs, they must enlist technicians trained in performing such sampling. Thus another key aspect of quality control is training workers, checking their performance, and rewarding them for high-quality work. The monitoring plan should provide quantitative thresholds for acceptable field measurements, with zero error the usual standard. The plan should also specify thresholds for further action. For example, if a project processes manure in an anaerobic digester, the monitoring plan could require quantification of emissions from manure stored more than seven days. For another example, if the value of soil organic carbon used in modeling is found to differ from the true value by more than 15 percent, the plan might require the modeler to redo the calculation.

Quantifiers would do well to gather the data they need to calculate baseline emissions when they are writing the monitoring plan, to ensure that those data are available. However, if the data are not available, quantifiers may simply detail the methods for calculating the baseline without actually performing the calculations. The monitoring plan should specify conditions under which quantifiers must recalculate the baseline to account for unexpected changes in conditions and should ensure that enough data are available to support recalculation. If quantifiers use a new version of a model to calculate emissions, they should use that same version to recalculate baseline emissions.

After drafting the plan, quantifiers should estimate the costs of completing the work and determine that enough funding is available. Verifiers acting as auditors will later use the plan to ensure that developers have implemented project activities and that quantifiers have taken the steps needed to calculate emissions reliably.

offsets because of unaddressed emissions, or because the target trading system will not recognize the offsets, verifiers should make this problem clear to the party seeking quantification.

Checking for Conflict of Interest

To provide legitimacy, verifiers must be independent of project developers, landowners, and offset buyers, and must never seek to maximize or minimize offsets. If the project developer hires a verifier to quantify offsets, there is a potential conflict of interest, although that might be acceptable if the relationship is disclosed. Having the project developer and offset buyer jointly select and pay the verifier reduces the likelihood of bias and broadens the verifier's obligations.

If a verifier has a continuing business relationship with a project developer (such as by participating in the design, selection, and marketing of projects), that person should not be verifying that developer's projects. Tests for conflicts of interest include the following:

Does the party have a direct (usually financial) interest in exaggerating the amount of GHG offsets? This is almost always presumed of the project developer, but profit-sharing mechanisms may also direct offset benefits to other parties. Verifiers' compensation should not be based on the amount of offsets they find.

Does the party have an indirect interest in biasing offset counts? An example of someone with such an interest would be an auditor who is also a consultant to the project developer. The auditor may reap financial rewards from the developer's own financial health through more lucrative consulting contracts, and thus he or she may have an incentive to inflate offset amounts. Indirect interest can also be personal rather than financial, such as family ties.

Does the party have an interest in showing that a project has achieved forecasted results? If a verifier projects future offsets and later becomes an independent verifier of the offsets, he or she would be motivated to show that the projections were accurate. Verifiers should disclose or rule out any potential conflicts of interest in their reports. If significant conflicts of interest exist, verifiers should decline the job.

Accrediting Verifiers

GHG mitigation systems may establish a list of approved verifiers based on an accreditation process and requirements (see Appendix 25). Systems typically accredit organizations, not individuals. That is because verifiers need the institutional durability and resources to assume liability for their opinions and attestations and because the broad range of skills needed for verification is better provided by a team than an individual.

Registering Offsets

If GHG offsets are to be fungible (fully transferable within a market framework) each offset must be listed in a registry that can credibly track the offset as it enters, resides in, and exits the market. In this way, the regulatory agency (if it exists) and the purchaser can be assured that the GHG benefits from a project have been uniquely assigned to the offsets being marketed.

Before starting verification, the project developer and verifier need to target a registry so they can tailor their work to its requirements. Though each registry establishes its own accounting guidelines, most are based on one of two principal sources. The World Business Council for Sustainable Development and the World Resources Institute have developed the *WBCSD/WRI GHG Protocol*, which nearly every registry that supports corporate reporting has adopted. The Intergovernmental Panel on Climate Change (IPCC) publishes guidelines for GHG accounting in geographically defined regions. The core IPCC guide is the *Revised 1996 IPCC Guidelines for National Greenhouse Gas Inventories*. Other IPCC guides include *Good Practice Guidance and Uncertainty Management in National Greenhouse Gas Inventories* and *Good Practice Guidance for Land Use, Land-Use Change and Forestry*. A registry is fundamentally a collection of information; even the largest could be stored on a single personal computer. Each record of offsets should include some or all of the following:

Serial number: A unique number associated with the record.

Offset size: The amount of GHGs sequestered or

avoided. Most GHG registries define an offset as one metric ton of CO_2e.

Offset type: An indication of whether the record represents an emissions allowance or offset, and possibly what type of offset[2] (that is, the method by which the offset is generated).

Vintage: The time period during which an allowance is valid or emissions have been reduced.

Account number: A code identifying the account to which the record belongs.

Project identifier: A code identifying the project that generated the offset.

Status: An indication of whether the offset is still available for trading or has been retired or canceled.

Project developer: The name and contact information of the entity that originated the offset (regardless of who owns the offset at the current time).

Geographic locator: Preferably an explicit geographic location of origin, or, at minimum, the geographic area in which the offset was created.

Links to the verification report: Unique identifiers locating validation and verification reports and supporting documents. These reports should be backed up off-site.

The registry monitors the status of offsets and allowances by requiring details documenting transactions and by maintaining a record of all transfers into and out of the registry and among participants (see Table 11.1). Registries often require both buyer and seller to notify the registry of any transfer of offsets, although some registries allow transfers based on a request from the initiating account only. The transaction record does not include the amount paid by the buyer, but the registry may play a monitoring role.

In a registry, *retired* offsets have been permanently removed from trading. For example, in the Kyoto system, at the end of the 2008–2012 compliance period, each country will retire allowances and offsets equal to its emissions during the period. This ensures that countries do not apply allowances to later emissions, and it also ensures the integrity of any emissions cuts.

A voluntary registry can retire verified offsets. The Climate Trust in Oregon, though lacking some of the formal structure of a registry, accepts GHG mitigation funds from customers wishing to offset their emissions. The trust uses the money to fund, validate, and retire GHG offset projects, taking into account additionality, leakage, and the permanence of the offsets. The buyer never actually takes ownership of the offsets; instead, he or she pays the trust to ensure that no other party can use them.

Some capped markets allow anyone to establish an account in their registry, purchase allowances or offsets, and retire them, pushing environmental protection above the mandated level. For instance, several individuals and nonprofits have bought sulfur dioxide allowances from the EPA's Acid Rain Program, thereby reducing total emissions.

Registered offsets are canceled—that is, disallowed and nontradable—when they are shown to be (1) unsupported by actual sinks or emissions cuts, (2) nonadditional, (3) of improper vintage, (4) registered elsewhere, or (5) otherwise in violation of registry or market standards. (The Kyoto system uses the term *canceled* differently. The units are considered canceled rather than retired when participants surrender allowances or offsets to meet compliance obligations or when entities without emissions caps acquire allowances or offsets and want to remove them from use.)

Ensuring Market Integrity

Ideally, before accepting offsets, each registry should ensure that they do not appear in any other registry. However, no registries now police against multiple registration, probably because the number of

Table 11.1 Transaction Record, Modeled on the Kyoto System

Transaction serial number	Offset list	Initiating account	Acquiring account	Date and time stamp
479432	789312, 789313, 789314, . . .	CH 001	AU 003	03/03/23 14:32:00

Note: The time stamp occurs after the transfer has been executed and validated.

markets is small and double registration would be obvious, but also because regulators usually do not have jurisdiction over other markets. At a minimum, registries should require parties registering offsets to attest that they are not registered elsewhere.

Ensuring intermarket integrity requires more than preventing multiple registration. The quality and value of offsets can vary from one registry to another, depending on the market rules. For example, the price of each ton of CO_2e can vary by a factor of more than 10, depending on the stringency of a market's emissions targets, a market's definition of baseline emissions, and participants' obligation to maintain offsets over time. Participants must take a close look at the market rules in which a registry is operating, not just the rules at the registry itself, before deciding to purchase a given offset. The adage "you get what you pay for" can apply to offset purchases.

Registering Reversible Offsets

Offsets involving carbon sequestration are by nature reversible. A large fire can immediately release thousands of tons of carbon sequestered over years of careful forest management and kill trees that will decay over time, generating further losses. In a properly functioning market, reversible offsets can have monitoring costs and risks that may lower their market value.

A reversible offset can be retired in two ways: it can be monitored in perpetuity and replaced if lost or it can be replaced with an irreversible offset after a specified length of time. A registry could identify reversible offsets as temporary. In other words, land-management offsets might remain valid only while monitoring shows that sequestered carbon continues to be stored and that the baseline remains valid. When the baseline expires, monitoring ceases, or a project ends for any other reason, the offsets expire. The expiring amount then counts as an emission in the account of the party that retired it.

When a Registrant Dissolves

Regulated markets do not allow registrants to withdraw from a registry, but voluntary markets cannot enforce such restrictions. However, a well-organized vol-

untary registry will establish a procedure for handling registrants' withdrawal. The best practice is to cancel offsets owned by a dissolved registrant or a registrant in default.

Voluntary registries can make two exceptions. In the first case, a registry may post units previously owned by a withdrawn registrant for sale. Such a sale must have a deadline tight enough to prevent the offsets from degrading. The second case occurs if the registry monitors the offsets itself, either while the offsets are awaiting sale or are retired.

Allowance-based registries can exert the most powerful environmental effect by retiring the allowances of a dissolved or defaulting registrant. However, markets that cap emissions from a specific industry may reasonably auction or redistribute some or all of the allowances because the remaining parties can expect greater demand for their products when a competitor leaves the market.

Ensuring Permanence and Transparency

As with any accounting system, a registry cannot erase transactions; it can only reverse them. Once registered, the record of offsets remains in the registry. If the offsets are reversed, retired, or otherwise canceled, the historical record ensures integrity and allows for audits by third parties or a regulating agency. Thus a registry's records will grow over time. A well-structured registry includes separate retirement and cancellation accounts, even though both are ownerless (no one can sell or surrender the offsets in them again).

Registries must make an appropriate amount of information publicly accessible to ensure widespread buy-in to the regulatory system. The public will want to know where offsets have been generated, the date of their generation, and the activities that created them. For those who wish to investigate further, public documents should clearly describe the baseline scenario, quantification methods, and proportion of leakage. These descriptions need not reveal proprietary information, and claiming a need to protect such information is not an acceptable reason to hide it. Revealing more information gives buyers and other observers better grounds on which to judge whether claimed offsets are real.

Publicly available information should include:

–Full records of all retired offsets.
–The identities of active account holders (participants who wish to remain anonymous can allow trustees or brokers to hold offsets and perform trades).
–The total amount of tradable offsets in the system (divided among types and vintages if the registry makes these distinctions).
–Monthly data for each type and vintage of offsets if the registry allows trades, including trade volume and open, high, low, and closing prices of offsets.
–Links to validation, quantification, and verification reports for all offsets. Public information may include abbreviated versions of these reports.

If the registry supports trading, participants are likely to have password-protected access to additional information, including the real-time asking price of offsets and the total amount available. Participants also need to know the vintage and type of offsets and have full access to project records, including complete verification reports and other supporting documents. If a registry does not support trading, it will need a mechanism for providing this information to an exchange that does.

Registry participants may not have access to all this information for offsets that are not for sale. A registry's decision on whether to make it available has little bearing on the quality of the market, though too much transparency can repel participants concerned about privacy.

Parties may trade anonymously through brokers, as is standard practice with stock exchanges. However, project developers cannot remain anonymous because their identities are an essential aspect of any due diligence by potential buyers.

A strong registry will provide a safe channel through which an auditor or any other party may report fraud. No existing system addresses this issue explicitly now, but registries should begin creating clear procedures for handling such incidents.

Chapter 12

Conclusion: Putting These Guidelines into Practice

This guide is intended to help practitioners and policy-makers understand and pursue the opportunity to create GHG offsets from farms and forests. This is a complex subject, so the production of a book on the subject must necessarily involve difficult editorial decisions on which technical information to include, which to reference, and which to omit. Implicit throughout this volume is the need to balance the use of the best methods with their costs. If the transaction costs of these methods are too high, land managers will not use the methods; if the methods used to develop offsets are not sufficiently rigorous, buyers will not wish to purchase the offsets.

From the perspective of practitioners who will develop specific projects in the field, this guide provides a beginning route to scoping, designing, developing, implementing, verifying, and registering GHG offsets. Those tasks will require a range of players, including not only farm and forest managers and their technical consultants, but also an array of service providers ranging from carbon analysts to independent verifiers, from investors to local farm advisors.

To support field practitioners, Environmental Defense is facilitating demonstration projects that apply the recommendations in this guide. In 2004, when the effort to produce it began, Environmental Defense entered into several key partnerships with organizations representing land managers in agriculture and forestry. An agreement with the National Association of Conservation Districts (representing 3,000 rural communities nationwide) has helped create regional cooperative programs with agricultural organizations around the country. Many of these organizations have decided to develop their own capacity to serve as aggregators for individual farmers and smaller groups of land managers.

The demonstration projects include the following:

– In Idaho, time-sequenced reforestation projects will return previously cultivated lands to pine forests, with the resulting offsets aggregated by a Native American tribe.
– In Kansas, Missouri, Nebraska, and Iowa, several hundred farms are applying no-till practices to corn rotation lands and supplying feedstock for biofuels, with aggregation provided by a cooperative mill.
– In Louisiana, some one hundred producers are pursuing cover crops, no- and low-till farming, reforestation, and capture of methane from livestock, with aggregation provided by state conservation districts.
– In Michigan, hundreds and even thousands of small landholders are converting to no-till cropping and reforesting marginal agricultural lands, with aggregation provided by local conservation districts and a statewide association of conservation districts.
– In New York, a group of small landowners and dairies in the central and western region is pro-

ducing offsets by combining reforestation, no-till farming, methane capture from manure, buffer zones, and cover crops, with aggregation provided by local conservation districts and a regional resource conservation and development corporation.

–In Ohio, local conservation districts and a statewide agricultural commodities organization are aggregating offsets produced through the use of buffer zones and no-till farming, reforestation of cultivated lands, and revegetation of mining lands.

–In Oregon, Washington, and Idaho, over one hundred farms are producing offsets from no-till farming, not only sequestering carbon in soil but also lowering fuel and fertilizer use, with aggregation by the Pacific Northwest Direct Seed Association, a nongovernmental organization with mostly farmer-members.

–In Xinjiang province in China, farmers are using no-till farming, a provincial forest agency is reforesting arid lands, village authorities are overseeing the capture of methane from manure at the village level, and a provincial water authority is reducing emissions through better water and fertilizer management, with aggregation by local and regional government entities, the lead agency being the Xinjiang Environmental Protection Bureau.

Other sites and other states may join these efforts, expanding the diversity of land-use practices and locations.

The second phase of these cooperative projects involves marketing—and ultimately selling—the GHG offsets. Two key activities will enhance the value and marketability of the offsets. One is creating a prospectus for each project, used to cultivate potential investors. Another is applying the quantification and verification standards outlined in this volume to the offsets proposed for sale.

Fortunately, broad audiences of farmers and landowners, prospective buyers, investment groups, and policymakers have expressed keen interest in witnessing real transactions involving actual contracts and transparent definitions of GHG offsets. However, without a cap-and-trade carbon market, demand for offsets will continue to be thin. A transparent, credible cap-and-trade market that includes GHG offsets cannot emerge, though, without policies and regulations that facilitate it.

This guide is particularly relevant to efforts to create regulatory standards for terrestrial GHG offsets. When these policies and regulations are in place, they will spur businesses to identify the most cost-effective ways of reducing their GHG emissions, resulting in significant demand for GHG offsets. That, in turn, will enable U.S. farmers, foresters, and other land managers to help reduce global warming while enhancing the value of their operations and the quality of their local environments.

The results of the demonstration efforts noted above will undoubtedly reveal the need for revisions to this guide. Continuing scientific advances in our understanding of how ecosystems interact with the climate (see for example Gibbard et al. 2005, Keppler et al. 2006) will also likely necessitate revisions. Results from the demonstration projects will also help inform the policy community on the intricacies of structuring a GHG emission market to effectively use offsets created by changes in land management. Most importantly, the experiences will help secure the credibility—and show the practicality—of farm and forest offsets as important elements in solving our global warming dilemma.

Appendices

Appendix 1

Key Factors to Consider in Developing a Sampling Strategy

Four practices are critical to ensuring that developers can accurately determine their project's offsets while not incurring unacceptable costs. These practices are: (1) paired sampling, (2) random sampling, (3) choosing an adequate number of sampling sites, and (4) stratifying the sampling sites when appropriate.

Paired Sampling

Accurately quantifying a project's GHG benefit often requires accurately measuring small changes in carbon stocks over a large area multiple times. That, in turn, requires minimizing the uncertainty that arises from variations in sequestration from one point to another. A *paired sampling* strategy can address this challenge. Paired sampling entails measuring carbon stock at designated locations, remeasuring those stocks at a later time at the same locations using the same techniques, and analyzing the change. Quantifiers then average sequestration over the entire project area and determine the statistical uncertainty in the average based on the variation from site to site.

This approach assumes that the sampling strategy is extensive enough to ensure that the average truly represents the project area. The other three strategies address that challenge.

Random Sampling

Sampling will not accurately measure the amount of carbon a project sequesters unless the process of locating plots is unbiased. For example, if quantifiers for forest projects locate plots in "nice" patches of trees, they will bias estimates of tree growth—and thus carbon sequestration—upward. Given a choice, field crews will avoid establishing plots amid brush because it is hard to navigate, but that will also bias sampling. The main strategy for establishing plots to obtain unbiased measurements is to sample randomly. The simplest approach to this is pure random sampling, where quantifiers use GIS software to randomly select the location of each sampling site.

An alternative to pure random sampling is systematic sampling with a random start. Systematic sampling can be aligned or unaligned (see Figure A.1). In aligned systematic sampling, field crews locate plot centers at a fixed distance from each other along parallel lines that are separated by a fixed distance. This approach allows crews to move from one plot to the next relatively easily. However, quantifiers must ensure that plot spacing does not correlate with some pattern in the landscape. For example, if the distance between plots is the same as that from ridge to ridge, and all plots happen to end up near ridge tops, the sample would be biased. In that case, unaligned systematic sampling may be preferable. In that approach, project lands are divided into a regular pattern of grids, and the sampling plots are located randomly within those grids.

Systematic sampling with a random start is recommended for large contiguous blocks of fields with more than 10 sampling sites per block. Random sampling is recommended when land parcels enrolled in a project

aligned systematic   unaligned systematic

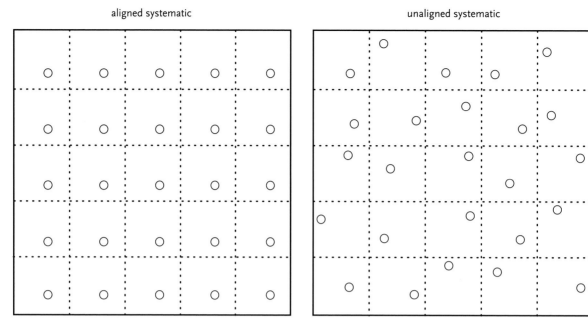

Figure A.1 Illustrations of systematic sampling of a project area using aligned and unaligned symmetric approaches. In both cases symmetric subplots are marked off—shown here as squares. In the aligned case, each sample is taken at the same place within the subplot; in the unaligned case, the samples are taken at random locations within the subplots.

are small, irregularly shaped, or narrow. Assigning plot locations before crews take the first field measurements allows either approach to work.

Choosing an Adequate Number of Sites

The number of sampling sites, combined with site-to-site variation, determines the statistical precision of a project's overall measurements. People often incorrectly assume that the number of sampling sites should reflect the size of the area being measured. In fact, spatial variability in carbon content across the project area—not the size of the area—should be the determining factor.

This spatial variability is characterized in terms of the *coefficient of variation (CV)*. *CV* is the average amount that the measured *change* in carbon stock on each plot varies from the mean change of all the plots. If *CV* is small, quantifiers can rely on a relatively small number of sampling plots to characterize the average amount of change in soil carbon over the project area.

If *CV* is large, they need to use a relatively large number of sampling sites.

In statistical terms, *CV* is the standard deviation of the mean divided by the mean or average.[1] For example, if measurements from all the sampling sites yield an average change in carbon stock of 2 tons per hectare, with a standard deviation of 0.4 tons per hectare, *CV* would be 0.4/2 = 0.20. According to convention, the coefficient of variation is expressed as a percentage of the mean: in this case, *CV* = 20 percent.

The most direct and accurate way to estimate *CV* for a given project area is to perform a pilot study that measures carbon stocks at various locations before the project begins. If a *CV* directly relevant to the project area is not available or attainable, analysts must rely on studies of other locations. However, many such studies report carbon stocks but do not cite variation in individual plots, and the sampling protocol in these studies may differ from those of the project. Still, estimating the coefficient of variation from existing measurements is better than merely guessing.

Once *CV* has been characterized, *n*, the number of sampling sites needed for a given level of accuracy and confidence, is given by

$$n = \left[\left(\frac{t}{e} \right) \times CV \right]^2 \qquad \text{Equation A1.1}$$

where *t* is the number of standard deviations (σ) needed to achieve the desired confidence level (typically obtained from a *t* table), and *e* is the allowable error (or uncertainty) in percent.

As Appendix 3 indicates, the change in carbon stocks should be accurate to within 10 percent, with a 90 percent confidence level using a one-tailed *t* test. Then, *t* = 1.3, *e* = 10 percent, and

$$n = \left[\frac{1.3}{10} \times CV \right]^2 = [0.13 \times CV]^2 \qquad \text{Equation A1.2}$$

If *CV* = 45 percent (a reasonable level of variability for most agricultural projects), then *n* = 35, a very tractable number of sampling sites. However, because *n* varies as the square of *CV*, the number of sampling sites can rise rapidly as the coefficient of variation increases.

For example, for a *CV* of 70 percent, *n* = 83, and for a *CV* of 100 percent, *n* = 169. Measuring changes in soil carbon on 169 plots is clearly much more costly than measuring changes on 35 plots. However, if the number of sampling sites is too small to provide accurate estimates of a project's offsets, then the project may not be profitable. (Recall from Chapter 2 that a project's claimed offsets are the calculated offsets minus statistical error or uncertainty.) If the coefficient of variation appears to be large, developers will have to decide whether to go forward with fewer sampling sites and lower accuracy, to absorb the cost of installing and measuring many plots, or to forego the project.

In fact, prudent developers will install more than the minimum number of sampling sites. The world is an uncertain place, and some plots may be lost as land is dropped from the project for any number of reasons. If a project will run for more than a few years, developers should install at least 15 percent more sites than the minimum so the possible loss of some sites will not undermine the entire project. If, as the project proceeds, quantifiers do not need the extra sites, they can drop some to save money.

Stratifying the Project Area

A common criticism of random sampling is that it does not guarantee complete "coverage" of important subpopulations. By chance, a random sampling design may not designate plots on every type of land, such as both forested and cultivated land. Stratified sampling—grouping similar areas expected to see similar amounts of change in carbon stock together and quantifying carbon in each group (see Figure A.2)—can guarantee that sampling represents all important

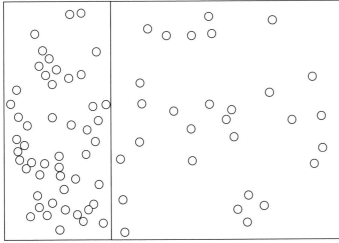

forested   cultivated

Figure A.2 Example of stratified random sampling. A common criticism of random sampling is that it does not guarantee complete "coverage" of important subpopulations. There is a finite chance that a random sampling design may fail to locate plots on every type of land: for example, forested areas within a largely cultivated tract. Stratified sampling—grouping together similar areas expected to see similar amounts of change in carbon stock—and quantifying carbon in each group can guarantee that all land-types are sampled. In the illustration, a simple random sample of *n* = 50 was selected in the forested (left) portion of the land holding; a simple random sample of *n* = 30 was selected in the cultivated (right) portion.

subpopulations. More importantly, if different areas are expected to register different amounts of change (forests are expected to store more carbon than cultivated areas, for example), stratified sampling usually produces more precise estimates by removing stratum-to-stratum variation in mean outcome from the overall estimate of standard error.

Quantifiers should base selection of strata boundaries on biological reasoning, rather than simply combining lands in a way that produces the maximum amount of sequestration. That means that quantifiers should group sites with similar amounts of change in carbon stock, even if the total amount of carbon in those groups varies. For example, consider a situation where the carbon stock in area A is expected to rise from 0 to 30 tons per hectare, the carbon stock in area B is expected to rise from 0 to 10 tons, and the carbon stock in area C to rise from 100 to 130 tons. Areas A and B have the same starting carbon stock but different expected amounts of change. Areas A and C have very different carbon stocks but are expected to see the same amount of change. Quantifiers should group areas A and C into the same stratum and place area B in a different stratum. The average of the carbon sequestration rate in each individual stratum is then determined independently and the total for the project is determined from the weighted average of the strata. If the variability in sequestration rates across the project lands can be grouped by strata, the estimate of the mean for each stratum will have a smaller uncertainty. Subsequently, the estimate of the total sequestration across the project area, using a weighted mean, will also have a lower uncertainty.

To determine how to stratify a project area, project developers can also use models and their knowledge of carbon dynamics. It is also acceptable to decide whether and how to stratify sites after fieldwork is complete, but care needs to be taken in this case to make sure that the stratification imposed on the project area does not introduce a sampling bias into the system.

Appendix 2

Quantifying Inadvertent Emissions from Project Activities

The activities that project managers pursue to sequester carbon usually create some CO_2 emissions, and they may change emissions of methane and nitrous oxide. The use of fossil fuel in managing project lands can produce the former, for example, while the use of fertilizer and changes in irrigation and tilling practices can affect the latter. In addition, all three gases are emitted during fires. While methane and nitrous oxide emissions may be relatively modest, they can still be important because their GWPs are considerably larger than the GWP of CO_2 (see Chapter 2). When these emissions occur on project lands or are caused by project activities, quantifiers must account for the net change in emissions, usually by using relatively straightforward methods.

Carbon Dioxide Emissions

Inadvertent CO_2 emissions most often occur from fossil fuel use and from fires.

Fuel use: CO_2 emissions from the use of fuel in project activities are usually small relative to the amount of carbon sequestered in wood—typically 1 to 5 percent of that amount. Project developers can therefore simply keep track of the types and amounts of fuel used during project operations and then use standard conversion factors to transform this information into CO_2 emissions (see Table A.1).

Fires: Landowners sometimes use fire as a management tool, especially in forest projects. Wildfires also sometimes break out on project lands, despite the best management practices. These fires produce CO_2 emissions, and quantifiers may or may not have to account for them, depending on the project.

Carbon sequestration projects usually do not have to track such emissions separately because fires deplete carbon stocks on project lands, and any measurement of biomass carbon stocks will therefore account for them. Biomass measurements made at or after the start of the project and before the fire indicate the amount of fine and coarse fuel. A fire expert can observe the site after a fire and estimate the amount of biomass remaining. Fire management agencies may also have models or measurements that estimate the amount of material burned in wildfires within the project area. Although estimating the area burned can be problematic, espe-

Table A.1 CO_2 Emissions from Burning Fossil Fuel

Fuel	Pounds of CO_2 per gallon	Metric tons of CO_2 per gallon
Gasoline	19.564	0.0088741
Diesel	22.384	0.0101533
Aviation gasoline	18.355	0.0083257
Kerosene	21.537	0.0097691

Source: U.S. Department of Energy, Energy Information Administration (http://www.eia.doe.gov).

cially when fire is patchy, a fair amount of uncertainty in the tons of dry matter burned is acceptable because emissions should be modest. Quantifiers can use whatever information they obtain from a moderate amount of investigation and use the default conversion factor of 94 percent to estimate emissions. If these emissions are less than 5 percent of net carbon sequestration, such estimates are adequate. If these emissions are larger than 5 percent of net sequestration, quantifiers should identify the rate of uncertainty in the estimate and use a high estimate of emissions or improve the reliability of the estimate.

Methane and Nitrous Oxide Emissions

Methane (CH_4) and nitrous oxide (N_2O) are greenhouse gases with GWPs of about 20 and 300, respectively, so quantifiers must take them into account when assessing a project's net greenhouse benefit. Both gases can be emitted from the burning of biomass and from soils and decaying organic material (such as forest litterfall, mulch used on croplands, or manure stored in a lagoon). Small amounts are also emitted when fossil fuels are burned, but quantifiers do not have to account for these emissions because the global warming impact is small. (Nitrous oxide emissions from vehicles are very small. Even multiplying by the high GWP of nitrous oxide, nitrous oxide emissions from vehicles are esitmated to have a net global warming impact of about 2 percent of that of CO_2 emissions from these sources, methane emissions on the order of 0.3 percent.[1])

While quantifiers must often account for methane and nitrous oxide emissions, reliably measuring them is a challenging task that can require significant human and monetary resources. For example, measuring methane and nitrous oxide emissions over a one-year period from even one field can cost $50,000 to $150,000—far more than the value of offsets most projects can generate. Quantifiers should therefore rely on model simulations, conversion factors, and empirical equations to estimate these fluxes (see the next section and Chapter 10).

Fires: Quantifiers can estimate methane emissions from fires as a function of their combustion efficiencies,[2] using the factors in Table A.2. Two fundamental

points are clear. Smoldering, smoky fires—those with combustion efficiencies of 0.9 or less—produce much more methane than hot, nonsmoky fires. However, the methane-related warming effect even from smoky fires is small relative to the warming potential of CO_2 emissions from burned vegetation.

For example, for a combustion efficiency of 0.88, the warming effect of methane released by the fire is only about 5 percent of the warming effect of the CO_2 released by that fire. In hot, low-smoke fires, the warming effect of the methane is less than 2 percent of the warming effect of the CO_2. Because fires tend to burn in a patch pattern, with different intensities at different locations, quantifiers will probably find more

Table A.2 Methane Emissions from Fire, by Combustion Efficiency

Combustion efficiency	CH_4 emitted (Mg CH_4/Mg dry matter burned)	Warming effect of CH_4 emitted as a proportion of the warming effect of the CO_2 from combustion (CH_4 GWP = 23) (Mg CO_2e/Mg CO_2e in biomass)
0.88	0.0042	0.06
0.9	0.0034	0.047
0.91	0.0030	0.041
0.92	0.0026	0.035
0.93	0.0023	0.031
0.94	0.0019	0.025
0.95	0.0015	0.020
0.96	0.0011	0.014

Notes: Mg = megagrams, or metric tons. GWP = global warming potential. The CO_2e of emitted CH_4 is given as a proportion of the total CO_2 emitted by the fire. The CO_2e of emitted CH_4 as a proportion of the CO_2 removed from the atmosphere and sequestered as carbon in biomass would be slightly lower. Combustion efficiency is defined in note 2. Note that fires may have lower combustion efficiency and thus higher methane emissions.

Source: The amount of CH_4 emitted as a function of biomass is from IPCC (2000), citing Ward et al. (1996).

Table A.3 Nitrous Oxide Emitted per Ton of Dry Matter Burned, as a Function of Nitrogen-to-Carbon Ratio in the Dry Matter

N:C ratio in %	kg N_2O/ton of dry matter burned
0.4	0.04284
0.6	0.05406
0.8	0.06528
1	0.0765
1.2	0.08772
1.4	0.09894
1.6	0.11016
1.8	0.12138
2	0.1326
2.5	0.16065
3	0.1887
3.5	0.21675
4	0.2448

Note: These calculations assume 1.7 metric tons of CO_2 emitted per ton of dry matter burned (in this case, different kinds of tropical vegetation).
Source: Equation 2.2.1 proposed by IPCC (2000).

than 5 percent uncertainty in the total amount of biomass burned. With this level of uncertainty, the entire warming effect of the methane emissions would be less than the uncertainty in the CO_2 emissions. In projects that sample biomass comprehensively, biomass measurements should capture the impact of the CO_2 emissions. In these cases, a comprehensive estimate of GHG emissions and sinks should include methane emissions from fire.

Burning vegetation also produces nitrous oxide. The Intergovernmental Panel on Climate Change (IPCC) suggests a default emissions rate that is a linear function of the amount of carbon burned and the ratio of nitrogen to carbon in burned biomass. (The IPCC did not formally adopt this method.) The IPCC equation for calculating the mass of N_2O emitted is

$$Ratio = 0.000012 + (0.000033 \times (N{:}C \text{ ratio}))$$

Equation A2.1

where *Ratio* is the number of tons of N_2O emitted per ton of CO_2 emitted, and *N:C ratio* is the mass of nitrogen in the burned biomass divided by the mass of carbon in that biomass times 100 (that is, expressed as a percentage).

If they lack site-specific data, quantifiers can assume that a ton of dry matter burned emits 1.7 tons of CO_2. Table A.3 provides examples of nitrous oxide emissions from different types of tropical vegetation. For all the types of forest tested, the warming effect of the nitrous oxide was less than 2 percent of the warming effect of the CO_2 emitted from the fire.

Soil and decaying organic matter: Chapter 9 provides methods for estimating these emissions and suggests land-management practices that can minimize them, for which project developers may receive offset credit.

Appendix 3

Using Statistics in Quantifying Offsets

Simply put, statistics help quantifiers design an effective strategy for assessing greenhouse gas benefits and (once sampling is complete) for calculating the amount of offsets and the uncertainty in that result. Consider a forest project where the quantifier needs to know the average diameter at breast height of all the trees. One option is to measure every tree. However, that approach is not practical for a large forest. Another option is to sample a subset of the trees and infer the average for the entire population from the the sampled subset. But how many trees does the quantifier need to sample to get a satisfactory answer—that is, an average that is acceptably close to the real average? And after conducting the sampling and calculating the average, how do quantifiers know how close the resulting average is to the actual average? Statistics help answer these questions.

The methods described here apply to all types of terrestrial GHG offset projects that use sampling rather than modeling. Quantifiers can rely on these relatively straightforward equations to calculate both the amount of offsets a project produces and the uncertainty in that result.

Definitions

This section describes populations, samples, and kinds of dispersion across samples. An analysis of these attributes shows how likely it is that a set of sample observations accurately reflects the population from which the samples are drawn. Table A.4 provides a list of the various parameters used to carry out this analysis, and Table A.5 contains a hypothetical data set used to illustrate the application of the concepts discussed here.

Population

The *statistical population of measurements*, often referred to as simply the *population*, is the set of measurements of a certain feature of a finite collection of objects. For example, consider the height of trees in a forest stand (see Figure A.3). The *actual mean height* is not necessarily the same as the *mean height of the population*, since the latter is based on measurements of a subset of the total stand. The better the sampling strategy, the more representative the population will be of the actual forest stand, and the more likely the two means will be comparable.

Sample

A *sample* is a subset of the population, such as a set of measurements of a certain feature on a subset of the objects of interest. A sample may be measurements of the heights of 1 percent of the trees in a forest stand or the average amount of carbon in the top 10 centimeters of soil in a core from the center of each of 30 square-meter plots selected from the grid partition on the land unit (Figure A.4). Because a well-designed sample can provide information about the entire population, sur-

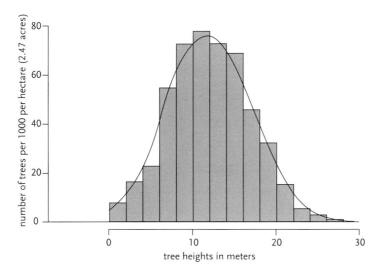

Figure A.3 The distribution of tree heights in a forest stand. The bars represent a histogram of measurements of a population or subset of trees measured. The solid line is an idealized representation of these measurements, assuming a Gaussian or bell-shaped distribution, referred to as continuous-density plot. Continuous-density plots are derived from the mean and dispersion—or standard deviation—in the population, assuming that the individuals in the population have a Gaussian distribution (follow a "bell curve"). For this discussion it is not necessary to understand exactly how the plot is derived.
Note: The information shows a mean of 12 meters, a standard deviation of 2.23 meters, and a variance of 5 square meters.

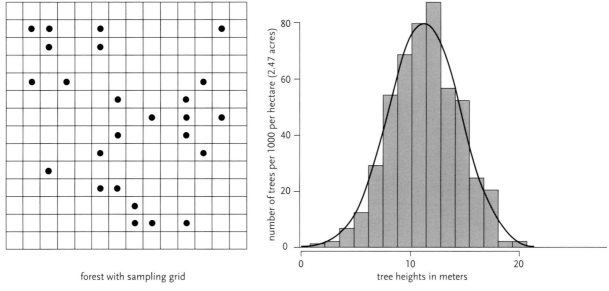

forest with sampling grid

Figure A.4 Hypothetical sampling of tree heights for the forest depicted in Figure A.3
(a) Forest outline with a 10 × 10 m² sampling grid, and the locations of 25 randomly selected plots that compose the sample marked by dots.
(b) Histogram of the sample of measured tree heights (normalized to thousands of trees per hectare), superimposed on the actual density of tree heights for the entire population. In this hypothetical example the random sample yields a mean of 13.5 meters, while the actual mean for the population is 12 meters.

veys can be designed to measure a sample of a population at a small fraction of the cost of measuring the entire population.

Note that the term *sample* denotes the measurements of a subset of objects chosen from the entire popu-lation. If the measurement of interest is obvious from the context, people often refer to the subset of objects (such as trees or square-meter plots) as the *sample* rather than the objects' measurements. This usage can cause confusion, even though it is common and used here on occasion as shorthand. Strictly speaking, the objects themselves are the *sampling units*, and the measurements of interest are the *sample observations*.

Another potential source of confusion is the distinction between the *sampling unit* per se and the physical sample manipulated by the measurement protocol. For example, a cultivated field may be considered a population of 1-square-meter sampling units, whereas the measurement protocol may simply call for taking a 6-centimeter-diameter soil core from the center of each sample unit. In terms of sample selection, total population size, and so forth, the sampling unit is a 1-square-meter plot, but in terms of actual measurements, the sample unit is the core.

The following examples often treat the plot and the core as equivalent. Treating the core as equivalent to the 1-square-meter plot assumes that the plot is homogeneous enough that the core accurately represents it. The distinction is mainly important in the initial stages of developing a measurement protocol, as it raises the question of *measurement scale*: how much of the sampling unit must be measured to achieve the desired accuracy?

Parameters and Statistics

The sample's *representativeness* determines the quality of the inference of a property of the population from the sample. Note that assessing representativeness requires first defining the characteristic(s) of interest in the population distribution: the center, dispersion, symmetry, general shape, and so on. Quantifiable population characteristics are termed *parameters* and denoted by lowercase Greek letters (see Table A.4). These are the actual and constant quantities that characterize the population that we want to estimate through sampling. The sample-based estimators of the actual parameters are called *statistics* and are denoted by Roman letters. For example, the estimator of the actual population mean, μ, is denoted by $\overline{y}$, and the estimator of deviation of values from the actual mean, σ, is denoted S. A key distinction is that a parameter is an unknown constant, whereas a statistic's value varies from sample to sample and hence is a random variable.

Common Characteristics and Their Associated Statistics

Center

The most common measure of central tendency is the *arithmetic mean* or *average*. The mean is a poor summary for skewed (i.e., nonsymmetric) distributions because it is heavily influenced by extreme observations. More robust measures for central tendency include (1) the *median*, the middle value of the sorted observations (for an odd population or sample size) or the average of the two middle observations (for an even population or sample size), and (2) the *trimmed mean*. A 10 percent trimmed mean is calculated by dropping the smallest 10 percent and the largest 10 percent of the observations and then calculating the mean of the remaining observations. The population mean is denoted by μ; the sample mean by $\overline{y}$.

Dispersion

The most common measure of dispersion is the *variance*. The definition of variance uses the mean, so this measure has limited interpretability for skewed distributions. The variance, a sum of squared values, is expressed in squared *units*. The square root of the variance, known as the *standard deviation*, is more commonly used since it is measured in the same units as the mean (see Table A.4). The population variance is denoted by σ^2, the population standard deviation by σ, the sample variance by S^2, and the sample standard deviation by S (sometimes lowercase s is used). To distinguish the standard deviation of a population from the standard deviation of a summary statistic, such as the standard deviation of the population and sample mean, the latter are denoted by $\sigma_{\overline{y}}$ and $S_{\overline{y}}$, respectively (see the section below on the standard error).

Coefficient of Variation

The *coefficient of variation* is a unitless measure of relative variation: the standard deviation divided by the mean. $CV = \sigma/\mu$ and $CV = S/\overline{y}$, for the population and the sample, respectively.

Table A.4 Common Distribution Characteristics, Associated Population Parameters, and Sample (Measured) Statistics

Characteristic	Parameter	Sample statistic	Example
Size; # of measurements	N	n	36
Center			
Mean	μ	$\bar{y} = \frac{1}{n}\sum_{i=1}^{n} y_i$	18.67
Median		$y_m = \begin{cases} y_{[m+1]} & n = 2m+1 \\ \frac{y_{[m]} + y_{[m+1]}}{2} & n = 2m \end{cases}$	18.32
10% Trimmed Mean	$\mu_{0.10}$	Drop the smallest 10% of the observations and largest 10%, then calculate the sample mean on the remaining observations.	18.52
Dispersion			
Variance	σ^2	$S^2 = \frac{1}{n-1}\sum_{i=1}^{n}(y_i - \bar{y})^2$	5.29
Standard Deviation	σ	$S = \sqrt{\frac{1}{n-1}\sum_{i=1}^{n}(y_i - \bar{y})^2}$	2.30
Coefficient of Variation (CV)	σ/μ	$S/\bar{y}$	0.123
Standard Error	$\sigma_y = \sigma/\sqrt{N}$	$S_y = S/\sqrt{n}$	0.38
Shape			
Skewness	γ	$G_1 = \dfrac{\frac{1}{n-1}\sum_{i=1}^{n}(y_i - \bar{y})^3}{S^3}$	1.163

Notes: The sample statistics assume a simple random sample design. The second-to-last column shows the sample statistics for the data on soil carbon (metric tons C ha^{-1}) in Table A.5. The sample statistic for the median differs depending on whether the sample size is odd, $n = 2m+1$, or even; the $y_{[m]}$ denotes the mth smallest observed value (the mth rank statistic).

Table A.5 Hypothetical Sample Data Used to Calculate Sample
Statistics in Last Column of Table A.4

15.53	19.55	19.95	20.05	18.19	17.63	22.68	18.46	26.84
18.83	18.88	20.19	18.04	17.29	20.47	18.82	21.98	18.64
20.58	14.32	17.48	15.69	18.14	17.59	17.74	16.73	16.62
15.18	19.88	20.65	17.84	20.22	17.94	18.88	17.32	17.39

Note: Each number represents a measurement of soil carbon (metric tons
C ha^{-1}) in 30 centimeters of core depth at one of 36 randomly selected plots
within a project area.

Standard Error

In an offset project, the mean value of a sample is typically used to represent the quantity in the population, such as the amount of carbon that a no-till project stores in soil or the amount of methane that a manure project captures. But because the mean value is based on measurements of a subset of the population, a quantifier must also determine the uncertainty or error in the sample mean—that is, how different $\bar{y}$ is from $\bar{\bar{\mu}}$. The standard deviation, S, is often mistakenly thought to be a measure of the error in the mean. That is not the case. The standard deviation simply tells us how widely the individual measurements in the sample vary around the mean value and thus how measurements vary from one random observation to another. It describes the distribution of values in the population and, at least in principle, does not depend on the sample size or strategy.

The uncertainty or error in $\bar{y}$, referred to as the *standard error* (as well as the *standard error of the estimate* or the *standard error of the mean*) and denoted by the symbol $S_{\bar{y}}$, is given by

$$S_{\bar{y}} = S / \sqrt{n} \qquad \text{Equation A3.1}$$

where S is the standard deviation and n is the sample size.

Thus as the population becomes more variable and S increases, the standard error of the mean also increases. That is, as the population measurements become more dispersed, using a limited set of measurements to accurately represent the population becomes more difficult. On the other hand, the error decreases if the sample size increases because the investigation is measuring a greater number of individuals.

Shape

The most important shape characteristic is *symmetry*, measured by *skewness* (see Figure A.5). Often, skewness is not quantified but simply noted from a graphical display of the observations. Positively skewed distributions have long right tails with a few very large values and a positive skewness parameter, $\gamma > 0$. Negatively skewed distributions have long left tails and $\gamma < 0$, and symmetric distributions have $\gamma = 0$.

Many commonly used statistics are reliable only if the distribution of the population is normal, which means that it is symmetric. Many populations, however, are skewed. For example, in the population of all the trees in the world, there are many more little trees than big trees. Fortunately, techniques for sampling and analysis allow the sample to be normally distributed even if the underlying population is not. This discussion assumes that the population and sample are close enough to normal distribution that the statistics in Table A.4 can be directly applied. (For a more detailed discussion of methods for addressing the problem of skewed distributions, consult an introductory or intermediate statistics textbook.)

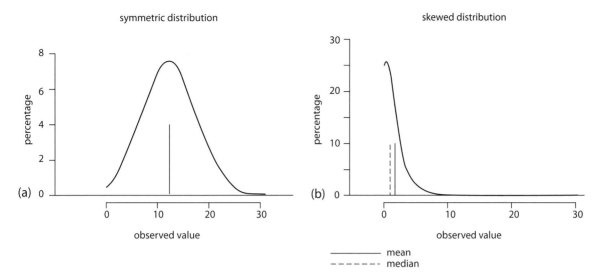

Figure A.5 Illustration of a normal (a) and skewed (b) distribution with skewness parameter $\gamma > 0$. Many populations can be skewed. For example, in the population of all the trees in the world, there are many more little trees than big trees and thus a plot of this population would tend to look more like (b) than (a). In this figure both plots are represented as idealized continuous-density plots, with no data shown for negative numbers. For each plot the solid vertical line marks the population mean, while the dashed line marks the population median. Because the normal distribution is symmetric, the mean and medium overlap.

Standard Error, Confidence Levels, and the Student t Test

The standard error, S_y, has a specific statistical significance: there is a 68.26 percent probability that the actual population mean, μ, is within one standard error of $\bar{y}$. That is, there is a 68.26 percent probability that

$$(\bar{y} - S_y) < \mu < (\bar{y} + S_y) \qquad \text{Equation A3.2}$$

There is also an 83.6 percent probability that

$$(\bar{y} - S_y) < \mu \qquad \text{Equation A3.3}$$

and that

$$(\bar{y} + S_y) > \mu \qquad \text{Equation A3.4}$$

Equation A3.2 is often referred to as a two-tailed test; the μ must lie within plus (+) and minus (–) one standard deviation of $\bar{y}$ —that is, it applies to both tails or sides of the distribution. Equations A3.3 and A3.4, on the other hand, are referred to as one-tailed tests because they apply to only one tail of the distribution.

Within this context, the probabilities of 68.26 percent and 83.6 percent are referred to as the two-tailed and one-tailed confidence levels of the standard error. Because the two-tailed test is more stringent than the one-tailed test, for a given confidence interval, the confidence level for the two-tailed test is less than that of the one-tailed test for the same value of the standard deviation.

A quantifier usually needs to know that offsets exist with at least a certain level of statistical confidence. This book recommends a 90% confidence level. The *Student's t test* is used to calculate the confidence interval, for a given level of statistical confidence:

$$\text{Confidence Interval} = t \times S_y \qquad \text{Equation A3.5}$$

where S_y is the standard error of the estimate and the value of t is a function the selected confidence level and the degrees of freedom of the sample (see Table A.6). The number of degrees of freedom is the sample size minus the number of variables. t values are typically denoted with the probability in the tail of the distribution. With this notation, $t_{0.10}$ has 10% of the probability in the tail of the distribution, and thus has 90% not in the tail. When performing a one-tailed test, this t value would give a 90% confidence interval. Subtracting the confidence interval from the mean estimated amount of offsets gives the amount of offsets for which there is a 90% level of statistical confidence that the true amount of offsets is at least the claimed amount.

As an illustration, consider the example used in Table A.5 of samples of soil carbon content on project lands. Recall that $\overline{y}$ = 18.7 metric tons C ha^{-1}, S_y = 0.38 metric tons C ha^{-1}, and n = 36. For a one-tailed confidence level of 90 percent, Table A.6 reveals that $t \sim 1.3$, and so

$$S_{y,90} = (1.3)\,(0.38\ \text{Mg C ha}^{-1}) =$$
$$0.49 \sim 0.5\ \text{metric tons C ha}^{-1}$$

In other words, there is a 90 percent probability that the actual soil carbon content of project lands is greater than or equal to 18.7−0.5 = 18.2 metric tons C ha^{-1}.

Because the major concern in offset projects is to avoid overstating greenhouse benefits, quantifiers can usually calculate the error in the mean using a one-tailed test instead of a two-tailed test. In most cases, a confidence level of 90 percent is best. And, as discussed in Appendix 1, the sampling strategy should be designed to achieve an error in the mean of 10 percent or less at the 90 percent confidence level. Appendix 1 shows how to do that using the statistical concepts covered here.

Table A.6: Student t Test: Values of t as a Function of Confidence Level, Sample Size (n), and Type of Test

Confidence level: one-tail test	25%	90%	95%	97.5%
Confidence level: two-tail test	50%	80%	90%	95%
Degrees of Freedom = n −1				
1	1.000	3.078	6.314	12.706
2	0.816	1.886	2.920	4.303
3	0.765	1.638	2.353	3.182
4	0.741	1.533	2.132	2.776
5	0.727	1.476	2.015	2.571
6	0.718	1.440	1.943	2.447
7	0.711	1.415	1.895	2.365
8	0.706	1.397	1.860	2.306
9	0.703	1.383	1.833	2.262
10	0.700	1.372	1.812	2.228
11	0.697	1.363	1.796	2.201
12	0.695	1.356	1.782	2.179
13	0.694	1.350	1.771	2.160
14	0.692	1.345	1.761	2.145
15	0.691	1.341	1.753	2.131
16	0.690	1.337	1.746	2.120
17	0.689	1.333	1.740	2.110
18	0.688	1.330	1.734	2.101
19	0.688	1.328	1.729	2.093
20	0.687	1.325	1.725	2.086
21	0.686	1.323	1.721	2.080
22	0.686	1.321	1.717	2.074
23	0.685	1.319	1.714	2.069
24	0.685	1.318	1.711	2.064
25	0.684	1.316	1.708	2.060
26	0.684	1.315	1.706	2.056
27	0.684	1.314	1.703	2.052
28	0.683	1.313	1.701	2.048
29	0.683	1.311	1.699	2.045
30	0.683	1.310	1.697	2.042
40	0.681	1.303	1.684	2.021
50	0.679	1.299	1.676	2.009
60	0.679	1.296	1.671	2.000
70	0.678	1.294	1.667	1.994
80	0.678	1.292	1.664	1.990
90	0.677	1.291	1.662	1.987
100	0.677	1.290	1.660	1.984
∞	0.674	1.282	1.645	1.960

Calculating Levelized
Costs and Benefits

Once developers and buyers have selected a discount rate and determined when a project's costs and benefits will occur, they can calculate the levelized cost of offsets. When costs and tons of offsets each accrue in a single period (but not necessarily in the same period), the levelized cost is

$$LevelizedCost = \frac{Cost/((1+d)^{n_c})}{Tons/((1+d)^{n_t})} \qquad \text{Equation A4.1}$$

where *Cost* and *Tons* are the undiscounted amounts, *d* is the periodic discount rate, n_c is the nth period in which costs occur, and n_t is the nth period in which tons accrue.

Consider the hypothetical projects cited in Chapter 4. One project would deliver 10 tons of offsets right now, for a payment of $100 right now. Because payment and delivery occur at the same time, the undiscounted and levelized prices per ton of offsets are the same: $10 per ton. Now compare this with the price per ton for a project that delivers 12 tons in 10 years, for a payment of $90 in year 3. The undiscounted price for this alternative is $7.50 per ton, which sounds attractive. However, now calculate the levelized cost, assuming a 6 percent annual discount rate and payments and deliveries that occur at the end of each year:

$$LevelizedCost = \frac{Cost/((1+d)^{n_c})}{Tons/((1+d)^{n_t})} = \frac{\$90/((1+0.06)^3)}{12/((1+0.06)^{10})} = \frac{\$75.57}{6.70} = \$11.28$$

The present value of the 12 tons delivered 10 years from now is only 6.7 tons, pushing the levelized cost to

$11.28—significantly higher than the cost of $10 per ton now to obtain offsets now. In this light, the alternative of waiting 10 years to receive the offsets looks considerably less attractive than simply creating or buying them now.

Analysts often use a discount rate of 4 percent per year to evaluate public investments. Using the standard 4 percent discount rate to levelize the same stream of costs and benefits yields

$$LevelizedCost = \frac{Cost/((1+d)^{n_c})}{Tons/((1+d)^{n_t})} = \frac{\$90/((1+0.04)^3)}{12/((1+0.04)^{10})} = \frac{\$80.01}{8.11} = \$9.87$$

In this example, changing the discount rate changes the rank of the alternatives. The levelized cost of offsets shrinks to $9.87 per ton—slightly less than the levelized cost per ton of the deal that occurs immediately. All other things being equal, investors would choose the option of waiting three years before making their payment.

Discounting Streams of Costs and Benefits

Projects will often yield offsets and incur costs over several years. In that situation, analysts can perform the calculations for each year and sum them:

$$LevelizedCost = \frac{\sum_{t=0}^{T} Cost_t/((1+d)^t)}{\sum_{t=0}^{T} Tons_t/((1+d)^t)} \qquad \text{Equation A4.2}$$

Table A.7 Hypothetical Offset Projects with Different Streams of Costs and Benefits (in Tons)

| Year | Project A | | | | Project B | | | |
	Cost	Tons	Discounted cost	Discounted tons	Cost	Tons	Discounted cost	Discounted tons
1	$1000		$961.54	0	$100	100	$96.15	96.15
2	$500		$462.28	0	$100	100	$92.46	92.46
3	$100		$88.90	0		100	0	88.9
4			0	0		40	0	34.19
5	$100		$82.19	0		40	0	32.88
6		100	0	79.03	$300	40	$237.09	31.61
7		100	0	75.99			0	0
8		100	0	73.07			0	0
9		100	0	70.26	$500		$351.29	0
10	$100	100	$67.56	67.56	$1000		$675.56	0
Total	$1800	500	$1662.50	365.91	$2000	420	$1452.60	376.19
Undiscounted $/ton	$3.60				$4.76			
Levelized $/ton			$4.54				$3.86	

Note: The table assumes a 4 percent annual discount rate, and that costs and tons accrue at the beginning of each year.

where $Cost_t$ and $Tons_t$ are the undiscounted amounts that occur in each period t, from period 0 to period T, and d is the periodic discount rate.

Consider an example where costs and tons for two projects accrue over several years (see Table A.7). As with many offset projects, the costs of Project A occur near the beginning, whereas most of the tons of offsets accrue later. Project B delivers offsets toward the beginning of the project (because project emissions converge with baseline emissions within a few years) while most of the costs occur later.

Project A delivers 500 tons of offsets for $1,800, which is an undiscounted cost of $3.60 per ton. Project B delivers 420 tons for $2,000, which is an undiscounted cost of $4.76 per ton. However, because of the timing of costs and delivery of offsets, the levelized cost is greater than the undiscounted cost for Project A, whereas the levelized cost is lower than the undiscounted cost per ton for Project B.

Spreadsheet software can easily determine levelized costs and benefits if analysts use one line for each accounting period. The discounted costs can be summed, the discounted benefits summed, and the levelized cost calculated.

Weighing the Present Value of Offset Rentals

Depending on the market price of offsets, a landowner may not be willing to commit to permanently maintaining offsets or replacing those offsets if they are lost through unplanned disturbances. However, the landowner may be willing to commit to creating and maintaining offsets for several years. In essence, temporary, or rented, offsets defer emissions until a later time. From the perspective of the buyer, or renter, the value is the cost of offsets now minus the present value of the cost to replace the expiring offsets in the future:

$$RentalValue_t = PriceNow - PVReplacement =$$

$$Price\ Now - \left(\frac{Replacement}{(1+d)^t} \right)$$

Equation A4.3

Where $RentalValue_t$ is the reasonable value of the rental of offsets for t periods, $Replacement$ is the price of acquiring permanent offsets at the end of the rental period, d is the periodic discount rate, and $PriceNow$ is the price of permanent offsets now. The rental value would also be adjusted for transaction costs and risk.

For illustration, assume that the price of permanent offsets is $10 and does not change, transaction costs are zero, there is no risk, the periodic interest rate is 4 percent per year, and the duration of the rental is 10 years:

$$RentalValue_{10} =$$

$$10 - \left(\frac{10}{(1+0.04)^{10}} \right) = 10 - \left(\frac{10}{1.48} \right) = 10 - 6.76 = \$3.24$$

As the interest rate rises and the term of the rental lengthens, the present value of replacing the offsets at the end of the rental term shrinks toward zero, and the value of the rental approaches the value of permanent offsets (see Table A.8).

Several factors can change the price a company will pay to rent offsets—beyond the expected cost of replacing them at the end of the rental period. For example, the renter will add the transaction costs of acquiring permanent offsets later. Buying offsets before they are created and renting reversible offsets are both more risky than buying irreversible offsets that already exist. If the expected periodic rate of failure equals the discount rate, the rental has no value.

The Impact of Changing Offset Prices on Rental Rates

These calculations assume that the cost of permanent offsets remains constant. However, a change in the price of offsets over time affects the value of a rental. Consider a situation where the cost of permanent offsets rises at the discount rate. Because the present value of the cost of future replacement offsets equals the price of permanent offsets today, the value of a rental is zero. Transaction costs would make the value of the rental negative.

If the price of permanent offsets is expected to rise faster than the discount rate, then the value of rentals is negative. That is because the present value of the cost of replacing the expiring offsets is greater than the cost of simply buying permanent offsets now. If the price of permanent offsets rises over time but at a rate slower than the discount rate, then this rise reduces the value of the rental but does not eliminate it. If the price of offsets falls over time, a rental allows a buyer to avoid acquiring expensive permanent offsets today and instead mitigate emissions with less expensive permanent offsets later. Thus falling prices of permanent offsets make rentals of offsets more valuable. A buyer will generally choose permanent offsets instead of rentals if it is cheaper to buy offsets or make internal emissions cuts. In no case will the price of a rental exceed the price of permanent offsets today.

Table A.8 The Value of Offsets Rented for Different Periods of Time

Term in years	2% discount rate	4% discount rate	6% discount rate	8% discount rate
5	9%	18%	25%	32%
10	18%	32%	44%	54%
15	26%	44%	58%	68%
20	33%	54%	69%	79%
25	39%	62%	77%	85%
30	45%	69%	83%	90%

Notes: This table provides rental values for different contract lengths and interest rates, where the renter pays the full value of the rental at the beginning of the rental period. If the number of offsets rented changes during the term of the rental, analysts would calculate value of the rental for each period, and discount those values, and then sum those discounted values.

Appendix 5

Categorical Additionality and Barrier Tests

In many carbon-trading systems, additionality is an either-or proposition. That is, a project is either deemed additional and can therefore proceed, or it is deemed not additional and cannot proceed. This type of system, known as *categorical* additionality, often relies on two criteria to judge additionality: commercial-use tests and barrier tests.

Commercial-Use Tests

Many carbon-trading systems consider projects as additional if the offsets they create will stem from "noncommercial" management practices. For example, if a project installs a manure digester to reduce methane emissions, and the only other digesters in the country are demonstration facilities financed through government grants, then the project is additional because farmers do not operate digesters commercially.

Different GHG mitigation systems set different thresholds for what counts as noncommercial. Most systems require that no more than 5 percent of installed capacity, or no more than 20 percent of new facilities and activities, use the project technology. The definition of "new" is usually about four or five years old.

Given the arbitrary nature of commercial use, how can regulators and offset retailers decide how best to employ such a test to determine additionality? A good rule of thumb is that the threshold for additionality should balance supply and demand for offsets so that their price is reasonable for buyers but they also pro-

vide a worthwhile return for landowners and project developers. With this approach, regulators would impose strict additionality rules—that is, a low threshold for defining what is commercial—in relatively small carbon-trading markets. In small markets, a loose additionality requirement would allow many offsets into the market that are not additional, driving down prices and making the system unprofitable for landowners and developers.

In large markets that aim to facilitate many GHG offsets, strict additionality rules can be counterproductive. That is because they will disqualify many projects with the potential to produce GHG benefits, inflating the price of the remaining offsets while also dampening efforts to mitigate emissions. New markets that anticipate significant growth might therefore choose strict rules for additionality in the beginning and then slowly relax those rules as the market grows. Avoiding climate change will ultimately require large cuts in GHG emissions, and that argues for large markets with relatively lenient additionality rules.

Table A.9 shows how additionality criteria might vary for small, medium, and large markets in a system that allows emitters to acquire offsets from anywhere in the world. Small markets need to use very restrictive criteria, such as categorical additionality, to avoid contaminating the supply of offsets with those that do not represent real GHG cuts. As the market grows, regulators can rely on more flexible criteria such as proportional additionality. Moderately large markets could allow multiple approaches to assessing additionality, before eliminating such tests when larger markets develop, to encourage the use of a wider variety of land-management practices. Very large markets might choose very lenient or even no additionality criteria to keep the price of offsets reasonable. If regulators aim to teach emitters to track and control emissions—and trade emissions cuts—in preparation for larger cuts, they should create rules that will apply to the later market.

Of course, even systems with no additionality standards must still require projects to establish baselines to ensure that the offsets they produce represent real GHG reductions. If projects must use baselines, they may produce real GHG benefits regardless of the criteria used to determine additionality.

Barrier Tests

Under the Kyoto Protocol's Clean Development Mechanism, the most common criterion for additionality are barrier tests.[1] With this approach, a project is considered additional if a barrier prevents most landowners from pursuing project activities or most landowners prefer other activities. Systems with a commercial-use test sometimes apply barrier tests to projects for which comparison sites are not available, to establish the extent of a land-management practice. Table A.10 lists barriers that might qualify a project as additional under this type of system.

In theory, a system using barrier tests would judge each project on its own merits and accept it as additional only if it faces one of the barriers. However, in practice, judgments about barriers often depend on access to proprietary information and assumptions about how the market will behave over the lifetime of the project. For example, trading systems often consider a lower economic return from a project activity than from an alternative land-use regime to be a barrier. However, showing that a lower economic return will actually occur from a project activity requires a detailed analysis of the assets and finances of the owners of any reference lands, as well as projections of the costs and revenues of project and non-project activities. Regulators usually have no way of checking landowners' assertions about their required rates of return, cash flow, and attitudes toward risk, and landowners usually guard such information closely to protect confidentiality.

More importantly, the results of that analysis can

Table A.9 Using Additionality Criteria to Manage Supply and Demand for Offsets

Market size, offset demand	Ratio between offset demand and supply	Additionality criteria
Small market (demand ~ 100s of millions of tons per year)	Demand for reductions is generally *less* than the potential supply.	Regulators adopt strict criteria. For example, categorical additionality requires that practices or technologies be in use on less than 5% of reference lands that create product(s) similar to the project's.
Medium market (demand ~ 1–10 billion tons per year)	Demand for reductions about *equals* potential supply.	Regulators adjust criteria periodically to reflect market fluctuations. For example: 1. When demand is relatively small, criteria are similar to those of small markets. 2. When demand begins to outstrip supply, proportional additionality or barrier tests take effect.
Large market (demand ~ 10–15 billion tons per year)	Demand for reductions *exceeds* potential supply by more than 8 to 1.	No additionality requirement, but baseline is still required.

Notes: This example assumes that emitters can acquire offsets worldwide. In more limited systems with fewer potential offsets, the definitions of small, medium, and large markets will vary accordingly.

Table A.10 Barriers to GHG Offset Projects That Could Be Used to Establish Additionality

Barrier category	Barrier description
Legal	– Unclear ownership rights – Poor enforcement of law, immature legal framework
Financial and budgetary	– High returns demanded by investors and lenders owing to poor risk/reward profile – Limited or no access to capital owing to poor functioning of capital markets
Technology, operation, and maintenance	– Higher perceived risks associated with new technology – Lack of trained personnel or technical and managerial expertise – Infrastructure changes required to integrate or maintain new technology – Inadequate supply of technology or transportation infrastructure
Market structure	– Price of outputs is below long-run marginal cost of inputs – Market distortions (subsidies or rules) favor another technology – High transaction costs

Note: This list is not exclusive.
Source: Adapted from World Resources Institute and World Business Council for Sustainable Development, Table 4 (2003).

be strongly influenced by its assumptions.[2] An analysis based on one set of reasonable assumptions may indicate a financial barrier, while an analysis based on another set of equally reasonable assumptions may not show a barrier.[3, 4] Because of these practical limitations, policymakers, quantifiers, and verifiers should use barrier tests as a last resort or to relax additionality criteria so offset projects can keep pace with demand.

When performing barrier tests, analysts should use information and assumptions that are as objective as possible, including third-party assessments of potential barriers. For example, the Organisation for Economic Cooperation and Development (OECD) issues Country Risk Classification ratings, which estimate the likelihood that a country will service its external debt (http://www.oecd.org), and Transparency International issues an index of corruption perceptions (http://www.transparency.org). These ratings independently indicate the risks that investors in offset projects are likely to perceive.

Appendix 6

Using Periodic Transition Rates to Calculate Baselines

Recall that to calculate the offsets credited to a project for each accounting period, quantifiers must determine the baseline for each accounting period. This is not a problem if project developers track land-use changes on comparison lands for each accounting period. However, such tracking may not be possible because of lack of access to comparison lands or long intervals between publication of land-use data. In that case, project analysts must infer such information from limited observations of management practices on comparison lands for each accounting period. Examples of specific types of projects illustrate this approach.

Afforestation Example: Deriving Periodic Transition Rates

Suppose a project converts marginal agricultural land to forest to sequester carbon in biomass. The project lasts 10 years, with 10 one-year accounting periods. At the start of the project, all comparison lands are used for marginal agriculture. New information does not become available until year 10, when it shows that 20 percent of comparison lands have been converted to forest, while the rest are still cropped. Calculating the baseline at year 10 is easy. If we assume that the marginal agricultural lands sequester no new carbon, and that all afforested acres gain the same amount of carbon at the same rate, the baseline at year 10 would simply be 20 percent of the project's sequestration on a per-acre basis. (See Equation 5.1 in Chapter 5.)

But what about the baseline during the intervening years? We cannot simply assume that the fraction of forested land in the project's fifth year is half that in year 10. The rate at which one type of land use transitions to another is usually proportional to the amount of land under that particular use:

$$FRL_i = fa_i \times TR_i \qquad \text{Equation A6.1}$$

where FRL_i is the fractional amount of land use type i that is converted to some other type over one accounting period (one year in our example), fa_i is the fractional amount of area in the comparison lands that is in use i during the accounting period, and TR_i is the so-called periodic transition rate from use i. With Equation 5.1 and observations of fa_i at any two points in time, the following equation can be used to derive the periodic transition rate:

$$FRL_i = 1 - \left(\left(\frac{fa_i^e}{fa_i^b} \right)^{1/(e-b)} \right) \qquad \text{Equation A6.2}$$

where fa_i^e and fa_i^b are the fractions of comparison land covered by land use i at e, the end of the land-use observation interval, and at b, the beginning of the land-use observation interval, and $(e-b)$ is the number of accounting periods (usually years) between the end and the beginning of the land-use observation interval.

It also follows that after m of the accounting periods (year m of the project),

$$fa_i^m = fa_i^b \times (1 - FRL_i)^{(m-b)} \qquad \text{Equation A6.3}$$

and the proportion of the comparison land area that is converted from use i to some other use is

$$Change_i^m = fa_i^b \times (1 - (1 - FRL_i)^{(m-b)}) \qquad \text{Equation A6.4}$$

To apply this methodology, analysts must have data on conditions on comparison lands for at least two different points in time. (Relying on more than two points makes the analysis more accurate but also more complicated. See below for restrictions on how long apart the two observations can be and when they can be made.)

In our afforestation example,

$$fa_{agric}^b = 1$$
$$fa_{agric}^e = 0.8$$
$$e - b = 10 \text{ years} - 1 \text{ year} = 9 \text{ years}$$

Substituting into Equation A6.3, we find that

$$FRL_{agric} = 1 - (0.8/1.0)^{1/9} = 0.024489$$

Calculating periodic transition rates by hand can be arduous, but it is quite simple with an electronic spreadsheet program. Table A.11 shows how to do this for our afforestation example. Recall that in this example, marginal agricultural land is assumed to yield no carbon sequestration, while forests sequester 0.5 tons per acre per year. The fractional coverage of each type of land use (agriculture and forest) is calculated for each year, using the equations in the text and a periodic transition rate from agriculture to forest of 0.024489.[1] The weighted benefit is the proportion of land in the particular use times the benefit per acre. B_A, the baseline, is the combined benefit from both types of land use (that is, the sum of the weighted benefits from agriculture and forests). The right-hand side of the table shows the baseline for each accounting period, as well as the accrued benefit as the project proceeds.

Further Complications

The simple example above assumed that the amount of carbon sequestered in forests is the same for all lands, regardless of how long they have been forested.

However, for afforestation as well as many other land-management practices, the greenhouse benefit per unit of input (such as an acre of land or a ton of manure) can change according to how long the unit has been subject to the practice. For example, tree seedlings typically sequester a negligible amount of carbon the first year after they are planted. However, in the tenth year, an acre of saplings may sequester as much as 3 tons of CO_2 equivalent. To account for such variation, analysts should consider the emissions from each tract of comparison lands for each accounting period. For instance, a tract that had been forested several accounting periods earlier would sequester more carbon than a tract that had just been converted to forest. The baseline is the sum of the sequestration on all tracts. These calculations sound complicated but are not difficult to perform with a spreadsheet program.

Tracts of comparison lands can also transition into more than one new management practice. For instance, in the afforestation example, some comparison lands initially managed as marginal agricultural plots could become pasture lands, while others could be newly subject to no-till cropping. In these cases, it is often easier to express the periodic transition rates for each type of transition as a matrix of probabilities. Such a matrix is often referred to as a *Markov* matrix, after the mathematician who developed this aspect of probability theory. For a readable presentation of such probability matrices, see Stokey and Zeckhauser 1978.

No-Till/Till Example: Calculating Baselines When Transitions Occur in More Than One Direction

A further complication arises when tracts of comparison lands cycle back and forth between two or more types of management practices. To establish the impact of such land-use changes on the baseline, analysts need to track them on each tract during the accounting period. It is not enough to simply track the total fraction of comparison lands devoted to each management practice.

Consider a project that moves from plowing to no-till cropping. Comparison lands consist of 20 one-acre tracts (see Figure A.6). During year 1 of the project, 4 of

Table A.11 Determining a Baseline for Each Accounting Period Using Periodic Transition Rates, Afforestation Example

| | Land-use type | | | | | | Combined benefit | |
| | Mariginal agriculture | | | Forest | | | | |
Year	fa_{agric} (Fractional coverage)	e_{agric} (C benefit in tons /acre)	Weighted C benefit (tons/acre)	fa_{forest} (Fractional coverage)	e_{forest} (C benefit in tons / acre)	Weighted C benefit (tons/acre)	B_A (tons/ acre) for each accounting period	B_A (tons/acre) accrued over project
1	1.00	0	0	0.00	0.5	0.00	0.00	0.00
2	0.98	0	0	0.02	0.5	0.01	0.01	0.01
3	0.95	0	0	0.05	0.5	0.02	0.02	0.04
4	0.93	0	0	0.07	0.5	0.04	0.04	0.07
5	0.91	0	0	0.09	0.5	0.05	0.05	0.12
6	0.88	0	0	0.12	0.5	0.06	0.06	0.18
7	0.86	0	0	0.14	0.5	0.07	0.07	0.25
8	0.84	0	0	0.16	0.5	0.08	0.08	0.33
9	0.82	0	0	0.18	0.5	0.09	0.09	0.42
10	0.80	0	0	0.20	0.5	0.10	0.10	0.52

B_A = Baseline emissions
Notes: Amounts are for illustration only and will vary widely on individual sites.
Source: Data from a survey of the literature reported by Smithwick et al. 2003, Appendix C, *Ecological Archives*, A012–012-A3, available at http://www.easpubs.org/archive/appl/A012/012/appendix-C.htm.

20 acres (or 20 percent) are in no-till cropping, and the remaining 16 acres (or 80 percent) are tilled. During year 10, 20 percent of the tracts are similarly in no-till cropping. Analysts might conclude that the transition rate from tillage to no-till cropping is zero. However, if the specific tracts that were in no-till cropping in year 1 are not the same as the tracts in no-till cropping in year 10, this conclusion would be incorrect. In fact, Figure A.6 reveals that only one of the four tracts in no-till cropping during year 1 remained in no-till cropping during year 10. Use of Equation A6.2 yields a transition rate from no-till to till cropping of 0.14. Of the 16 tracts plowed in year 1, three are not tilled in year 10. This yields a transition rate from till to no-till cropping of 0.023. The transition rates differ because the initial fractions of tilled and no-till lands in each accounting period are different, and the rate of transition is proportional to amount of land in that specific use (See Equation A6.1).

Table A.12 shows how analysts might use a spreadsheet program to calculate the baseline in this no-till/till example. Fractional coverages for no-till and tilled tracts must be calculated for each new tract as well as the remaining tracts. Note also that whereas 20 percent of tracts are in no-till cropping at both the start and end of the project, less than 20 percent are in no-till cropping in the intervening years. Thus the baseline is slightly lower than if analysts did not account for this transition.

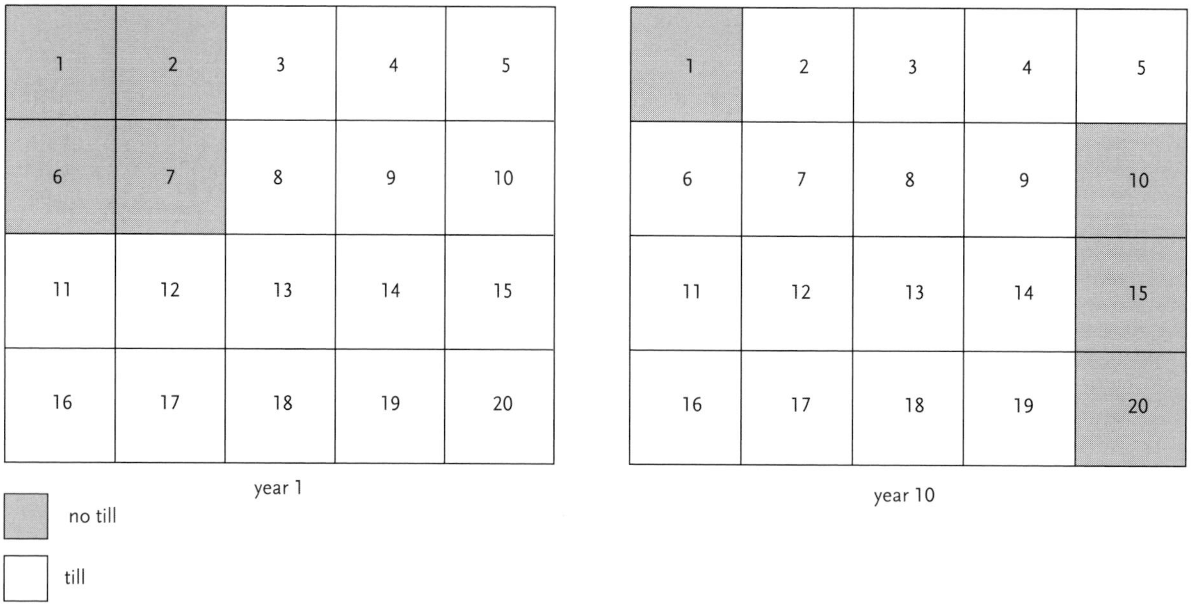

1	2	3	4	5
6	7	8	9	10
11	12	13	14	15
16	17	18	19	20

year 1

□ no till

□ till

Figure A.6 Land-management practices on hypothetical comparison lands used to establish the baseline for a no-till project. The comparison lands consist of 20 one-acre tracts, each indicated by a box. Management practices for each tract (grey for no-till and white for till) are illustrated for years 1 and 10 of the project. While years 1 and 10 have the same number of boxes in no-till, there is nevertheless a non-zero transition rate from no-till to till because the locations of the boxes have changed.

How Far Apart Can Land-Use Observations Be?

Measurements of land conditions made more than a decade apart are unlikely to be useful. On the other hand, measurements at least three to five years apart can smooth out year-to-year variability in GHG emissions and carbon sinks on comparison lands that may not be representative of the entire region. However, if a land-use sector is changing dramatically, analysts should base the rate of change on information gathered only two years apart. If the most recent information is at least five years old, analysts should assess whether the rates of change they observe are likely to be similar to current rates. If the first measurement occurred more than 10 years before the start of the project, significant changes in the economics of producing project goods can make the information unreliable.

Ideally, land-use observations are made close to—if not exactly coincident with—a project's accounting periods. Extrapolating transitions well beyond the last land-use observation is especially risky. The formulation used to describe transition rates is asymptotic; that is, those rates approach 0 for the proportion of land remaining in the original state, and 1 for the proportion of land in the new state. Thus, using the equations for time periods much longer than the actual observed interval can lead to results that are far from reality. Baselines with smaller periodic transition rates are less likely to be wrong by a given amount than baselines with larger rates. Moreover, baselines should not be considered reliable for more than 15 or 20 years without land-use measurements spanning the project period. If a project is scheduled to run longer than 20 years, the baseline should be recalculated using new land-use data when the project ends.

Table A.12 Calculating the Baseline with Tracts Cycling between No-Till and Till Cropping

| | Land-use type | | | | | | | | Combined benefit | |
| | Tilled tracts | | | | No-till tracts | | | | | |
Year	f_{till}-rem (fractional coverage of initial tilled tracts remaining)	f_{till}-new (fractional coverage of newly tilled tracts)	f_{till}-total (total fractional coverage of tilled tracts	Weighted benefit from tilled tracts (tons/acre)	$f_{no\text{-}till}$-rem (fractional coverage of initial no-till tracts remaining)	$f_{no\text{-}till}$-new (fractional coverage of new no-till tracts)	$f_{no\text{-}till}$-total (total fractional coverage of no-till tracts)	Weighted benefit from no-till tracts (tons/acre)	B_A (tons/acre) for each accounting period	B_A (tons/acre) accrued over project
1	0.80	0.00	0.80	0.08	0.20	0.00	0.20	0.10	0.18	0.18
2	0.78	0.03	0.81	0.08	0.17	0.02	0.19	0.09	0.18	0.36
3	0.76	0.05	0.82	0.08	0.15	0.04	0.18	0.09	0.17	0.53
4	0.75	0.07	0.82	0.08	0.13	0.05	0.18	0.09	0.17	0.70
5	0.73	0.09	0.82	0.08	0.11	0.07	0.18	0.09	0.17	0.87
6	0.71	0.11	0.82	0.08	0.09	0.09	0.18	0.09	0.17	1.04
7	0.70	0.12	0.82	0.08	0.08	0.10	0.18	0.09	0.17	1.22
8	0.68	0.13	0.81	0.08	0.07	0.12	0.19	0.09	0.17	1.39
9	0.67	0.14	0.81	0.08	0.06	0.13	0.19	0.10	0.18	1.57
10	0.65	0.15	0.80	0.08	0.05	0.15	0.20	0.10	0.18	1.75

B_A = Baseline emissions

Appendix 7

Typical Carbon Stocks
in Forest Pools

In this appendix, we review the types and characteristics of the various stocks or pools of carbon that must be inventoried in a forest project.

Live Trees

Tree biomass is usually the carbon pool with the largest potential for sequestration. On sites that have been recently clear-cut, grassland, and agricultural sites, live-tree biomass is essentially zero. In woodlands, because of wide tree spacing and small tree sizes, carbon stocks are generally not more than 20 to 40 metric tons of carbon per hectare (Mg C ha^{-1}). The exception is woodlands where herbivory, fire, or another condition limits tree establishment but available water and nutrients allow common tree species to grow to substantial size. Examples include some tropical acacia savannas and western U.S. ponderosa pine woodlands.

In forests managed for timber, carbon in live trees generally does not exceed 50 to 100 Mg C ha^{-1} (Birdsey 1996). In undisturbed mature temperate and tropical forests, typical carbon stocks are 100 to 200 Mg C ha^{-1} (Houghton 1999). Stocks can be much greater in some moist forests that grow very large trees. Live-tree carbon in 450-year-old Oregon Cascade Douglas fir forests was measured at 584 Mg C ha^{-1}, and live-tree carbon in 150-year-old Oregon coastal hemlock-spruce forests was measured at 626 Mg C ha^{-1} (Smithwick et al. 2002). The global maximum live-vegetation biomass probably occurs in old-growth stands of coastal redwood forest (*Sequoia sempervirens*) along the coast of northern California. One stand older than 1,000 years was calculated having a stem biomass of 3461 Mg C ha^{-1} (Waring and Franklin 1979).

Standing Dead Trees

Standing-dead-tree carbon varies tremendously as a function of climate, forest type, stand age, and disturbance history. In young, planted forests, and forests with aggressive thinning, carbon stocks in standing dead trees can be almost nil. In forests where wind is the primary disturbance agent, carbon stocks in standing dead trees will also be small. Examples of U.S. forest types where carbon in standing dead trees is often close to zero are spruce-fir forests in the Northeast, oak-pine forests in the northern plains, planted pine forests in the Southeast, and red alder forests in the Pacific Northwest (Smith, Heath, and Jenkins 2003).

In other cases, carbon stocks in standing dead trees can be large. Although standing-dead-tree carbon stocks of 0 to 20 Mg C ha^{-1} are typical in U.S. forests, stocks of up to 50 Mg C ha^{-1} are not extraordinary (Smith, Heath, and Jenkins 2003). After a wildfire, dead-tree biomass can be 90 percent or more of the live-tree biomass immediately before the fire. This is because fires almost never consume the boles of live trees. Live trees that experience even hot crown fires usually lose only their foliage, fine branches, and some of the bark on the trunk.

Woody Debris

Carbon stocks in coarse woody debris are highly variable across forest types (see Table A.13). In the majority of forest systems, most dead trees do not completely decompose standing up. Instead, they fall and become coarse woody debris. Some stands, however, have almost no such debris. Stands with low woody debris levels include young and mature stands afforested from open land, stands that have been intensively managed for several decades in a way that removes dead and dying trees, and stands with very fast decomposition rates.

Young and mature stands established on open land have no legacy of woody debris from a prior stand. In intensively thinned stands, and stands where dead and dying trees are often removed, wood that would have become coarse debris is harvested. Some species decompose quickly, and some forests have generally high decomposition rates. In some tropical forests, the

Table A.13 Typical Live and Detrital Carbon Stocks of Selected Vegetation Types

Vegetation type	Vegetation Mg C ha^{-1}	Detritus Mg C ha^{-1}
Tundra and alpine	3	216
Desert scrub	5	56
Temperate grassland	7	192
Temperate woodland	27	no data
Tropical woodland	60	69
Old-growth Ponderosa pine, eastern Oregon, USA	113	45
Boreal/subalpine	117	149
Temperate broadleaf deciduous	143	118
Mature tropical/subtropical	247	104
Old-growth Douglas-fir, Cascades, Oregon, USA	557	207
Old-growth spruce/ hemlock, coast, Oregon, USA	598	164

bulk of fallen tree biomass is shredded and is no longer coarse debris within three years after it falls to the forest floor.

Several factors can produce large stocks of coarse woody debris. Such stocks can be substantial immediately after harvest. Harmon, Garman, and Ferrell (1996) report woody residue rates after the harvest of U.S. Pacific Northwest mature and old-growth forest in the range of 36 to 52 Mg C ha^{-1} (assuming that biomass is half carbon). Forests where trees grow large and are not harvested (particularly where tree species have decay-resistant wood) can accumulate large stocks of woody debris. Krankina and Harmon (1994) report stocks in the range of 18 to 75 Mg C ha^{-1} in old-growth forests in the U.S. Pacific Northwest. Cool temperatures slow decomposition, and boreal and alpine forests can accumulate large stores of woody debris. Extreme wetness and extreme dryness also slow decomposition. Peat bogs are the extreme case: decay is slower than the rate of input, and vegetative material accumulates indefinitely. In dry forests, stocks of woody debris that does not burn can be substantial compared with stocks of live-tree biomass.

Litter

Studies of carbon on the forest floor often include both undecomposed litter and decomposed organic material above mineral soil. Litter is undecomposed and minimally decomposed vegetative material on the surface of the ground, including foliage, twigs, bark, and seed cases. Some forests have layers of partially decomposed duff or organic soil horizons composed largely of humified organic material. There is no standard definition of what assessments of carbon in litter and on the forest floor should include. Studies of litter and fine debris have used maximum piece diameters from 6 millimeters to 10 centimeters in diameter (Harmon and Sexton 1996). If measurements of carbon stock on the forest floor include duff and organic horizons, live roots above a specified size are removed. As with woody debris, there is no standard minimum size, but studies often remove live roots greater than 2 millimeters in diameter.

Carbon stocks in litter vary by time of year. In most

forests, between half and nearly all of annual litter input decomposes within a year. In many closed-canopy, seasonally deciduous forests, litter stock is close to zero at some point each year. The dynamics that drive stocks of litter and coarse woody debris overlap. If not disturbed for many decades, forests with slow decomposition rates can accumulate substantial carbon stocks in organic layers (Grier et al. 1981; Means, MacMillan, Cromack 1992).

Understory Vegetation and Shrubs

Carbon stocks in understory vegetation and shrubs are usually small, less than 3 Mg C ha^{-1} (Birdsey 1996, Smithwick et al. 2002). On sites where species classified as shrubs form the overstory, rather than the understory (and where these species grow in dense stands that are several meters tall, with stems several centimeters in diameter), carbon stocks can reach 30 Mg C ha^{-1} (G. Smith, unpublished data, 1999). If shrub species are measured separately from species that may grow into larger trees, then the shrub category can store moderate amounts of carbon.

Appendix 8

Protocols for Measuring Carbon in Subplots

The methods and protocols used to measure the carbon in subplots are covered here.

Trees

To begin biomass field measurements after marking plot boundaries, establish a subplot with a 2.52-meter radius around the plot center. Correct all horizontal measurements for slope if it is greater than 10 percent (see Appendix 11). Working clockwise from map north, for each stem where the center of the tree trunk at ground level is within the plot and the stem's diameter at breast height is at least 5 centimeters, record the following: species code, diameter, height, top diameter, and vigor/decay class. (Diameter at breast height, or dbh, is 1.37 meters above the forest floor on the uphill side of the tree.) Measure the top diameter of broken trees using the optical dendrometer. Include all live stems, snags, and stumps of the size class included in the subplot. Snags and stumps are defined as dead plants with woody stems still attached to their roots, where the roots are mostly in the ground, and where the growth axis is within 45 degrees of vertical (stumps are very short snags). When stumps are less than 1.37 meters tall, record the actual top diameter, not the estimated diameter at 1.37 meters of height.

Establish a subplot with a 5.46-meter radius centered on the plot center, slope-corrected as necessary. Working clockwise from map north, for each stem originating within the plot that is at least 15 cm dbh, record the following: species code, diameter, height, top diameter, and vigor/decay class. Include all live stems, snags, and stumps.

Establish a subplot with a 17.84-meter radius centered on the plot center, slope-corrected as necessary. Working clockwise from map north, for each stem originating within the plot that is at least 30 cm dbh, record the following: species code, diameter, height, top diameter, and vigor/decay class. Include all live stems, snags, and stumps.

Field crews may find it more efficient to record all live trees on a subplot of a given radius and then to record snags, especially if the plot includes many stumps or other short snags.

Litter

Destructive sampling requires special care to minimize the possible effects on later remeasurements. If sampling aims to determine project-specific factors, such as the densities or carbon content of dead wood, this sampling should occur outside the subplots used to measure changes in biomass. Within such subplots, destructive sampling should be minimal and limited to carbon pools that will not affect the change in carbon stock. For example, destructive sampling of tree seedlings should be avoided because it might reduce tree density later, thus reducing biomass. Nondestructive measurements are often quicker to make and thus less expensive.

When remeasuring biomass on plots with previous destructive sampling, crews may have to move the location of destructive sampling points. For example, sampling soil entails taking multiple cores from points in a specified arrangement around the plot center. During resampling, crews should take cores from points between the points initially sampled. However, if the sampled pool turns over almost completely between sampling times, destructive sampling may have little impact on remeasurements. For example, more than 90 percent of litter in many moist temperate systems decomposes in fewer than three years. If crews measure litter at the same locations only every five years, the previous measurements should have little impact.

The protocol in Chapter 6 calls for sampling litter from a 0.5-meter-by-0.5-meter subplot. All the litter from the subplot is gathered and weighed, and a representative subsample is selected and weighed in the field. The subsample is then taken and dried to yield a ratio that converts field weight to dry weight. Crews should redistribute the litter they do not remove across the subplot to minimize alteration of plot dynamics.

If the litter, microclimate, and weather are consistent from plot to plot, then crews do not have to subsample litter on every plot. However, if any of these influences does change, crews should subsample litter on all the plots because the ratio of dry weight to wet weight can change quickly and dramatically. Some decayed organic material can absorb more than four times its dry weight in water, or it can dry in the field to nearly dry weight. A rainstorm can change the dry-to-wet-weight ratio from greater than 0.85 to less than 0.2. Failing to obtain a plot-specific ratio can mean that the estimate of carbon stock in litter is off by a factor of four.

Chapter 6 recommends classifying all fine organic material on the ground above mineral soil—including loose, undecomposed leaves, twigs, bark, seeds, and other identifiable plant parts—as litter. Litter usually also includes duff: matted, partly decomposed plant parts that are still readily identifiable with the naked eye. Duff is the same category as the Oe organic soil horizon (see below).

Some plots have a layer of decayed organic mate-rial below the undecomposed litter and duff and above mineral soil. Soil scientists divide this dark-colored material into the Oe organic horizon of intermediately decomposed (hemic) material and the Oa organic horizon of highly decomposed (sapric) material. A bit of this decayed organic material rubbed between thumb and forefinger feels smooth. A bit of mineral soil—except for clay—feels gritty. Most sites do not have layers of decomposed organic material. If a significant amount of the Oa horizon exists within the project area, the sampling protocol should include a procedure for separately measuring this layer.

Careful fieldwork is essential. It is especially important to avoid including mineral soil when sampling litter. This hazard is less of an issue when crews are separating minimally decomposed material from an O horizon, but it is harder to avoid when decomposing chunks of wood are partly buried in mineral soil. Litter often is loose or forms a mat. In either case, field crews can usually lift it gently off the underlying soil, picking up stray pieces individually. Crews should clean the mineral soil of partially buried pieces of litter before packaging them with the rest. They can bang solid pieces against a tool to knock off the soil. Crews are reduced to brushing, picking, and blowing soil off less-than-solid pieces. Those pieces that are more than half-buried in mineral soil are classified as buried and sampled with the soil, not the litter or organic layer. Soil experts must teach technicians how to separate the O horizon from underlying mineral horizons.

If the locations of subplots change from one measurement time to the next, quantifiers must consider variability in these locations. For example, in some forest systems, substantially more litter collects in depressions that are a few centimeters deep and tens of centimeters across than on higher locations. Moving a 0.5-meter-by-0.5-meter litter subplot 1 meter sideways can significantly change the amount of litter that crews will find. However, this should not be a problem for forestry projects because litter is usually a small carbon pool, so increasing the variability of litter measurements will have a negligible effect on the precision of the overall measurement.

Disturbance can sometimes significantly change

the stock of decomposed organic material on the forest floor that is above mineral soil. This material generally accumulates slowly, and disturbance increases respiration, which can decrease the amount of carbon. On most sites, this disturbance is not a problem for repeated measurements because the total stock of decomposed organic material is small. However, on sites where decomposition rates are very slow because of nutrient limitations, pH, moisture, cool temperature, or some other reason, the stock of decomposed organic material can be high. On such sites, resampling should occur at locations that have not been previously sampled. To average out fine-scale variability, it may be desirable to use a coring system similar to that described in Chapter 7 for sampling soil, instead of a single subplot.

Woody Debris

Unless a site has extraordinarily slow decay rates, small pieces of woody material and decomposed organic material will never form a great mass, and an efficient sampling design will devote little time to measuring such material. One way to speed measurement is simply to count the number of pieces within a particular size class, rather than measuring the exact diameter or length of each piece. Quantifiers then use the median size to calculate the mass for each class. This approach is valid for fine debris and small live woody plants.

Counts of pieces within a size class proceed quickly if crews use calipers or size gauges. Gauges can be of fixed size, with one gauge for each class boundary. Pushing the gauge against a piece instantly reveals whether it is larger or smaller than the size limit (see Figure A.7). Gauges can be stepped, with smaller sizes in the throat of larger openings. Stepped gauges are pushed against the material until an opening is too small for the material to fit inside. Gauges can be made for the desired thresholds so there is little confusion as to whether the gauge in a field technician's hand is the desired size. Gauges can be fabricated from sheet metal thick enough to resist bending under the abuses of fieldwork. A string tethering the gauge to a technician's field vest reduces the chance that it will be lost or stepped on.

Alternative methods for measuring coarse woody debris appear in Harmon and Sexton (1996), Brown (1974), and van Wagner (1968). Another common method is to measure the length of each piece of coarse woody debris and its diameter at each end and then compute the volume as a section of a cone. However, this method is not recommended because it takes longer than the suggested method. If the transects used to measure woody debris are long enough to cross multiple pieces, and the sampling system measures debris along several dozen transects, the suggested approach is almost as accurate as the more labor-intensive method.

Harmon and Sexton (1996) recommend transects at least 100 meters long because spatial distribution of woody debris is patchy. Field crews can quickly establish a 100-meter transect by running lines 25 meters in each cardinal direction from the plot center. Even in stands with relatively small amounts of coarse woody debris, 100-meter transects yield a small standard error when only a few dozen plots are measured. Two short transects that meet at a right angle reduce possible sampling bias, especially if the orientation of coarse woody material is not random (van Wagner 1968). This is not just a theoretical problem. In forest types where wind is a common agent of tree mortality, the orientation of fallen tree trunks tends to be correlated. If a single transect is parallel to the main direction of tree fall,

Figure A.7 A 5-, 3-, and 1-centimeter caliper used to measure woody stems 1 to 2 centimeters in diameter. Counts of woody pieces can be expedited by using calipers or size gauges.

woody debris loads would be undercounted. If a single transect happens to be perpendicular to the main direction of tree fall, woody debris would be overcounted. Placing two transects at a right angle on each plot reduces this potential bias. In forests with consistent debris distribution, somewhat shorter transects may be acceptable.

Logs generally become elliptical in cross section as they decay. Field crews can measure the maximum and minimum diameters of such logs with minimal difficulty. Decomposed debris is typically resting on the ground, and it is not possible to get a diameter tape under the piece, however. Strictly speaking, diameter tapes overestimate the cross-sectional area of elliptical pieces, but when crews are measuring live trees, the actual error is small (Parker and Matney 1999). Using a 100-centimeter adjustable caliper enables crews to avoid parallax, which easily occurs when they hold a measuring tape in front of a piece to estimate its width. Technicians can minimize parallax if they sight along a line perpendicular to the distance they are measuring to align the zero point on a tape measure with one edge of the piece. Technicians then move their head to sight along another line parallel to the original line, with the second line tangential to the far edge of the piece, and find the diameter by reading the tape.

If crews measure the diameters of the ends of each piece of debris and piece length, they should treat each piece as if it ended at the plot boundary, or they will overestimate the amount of woody debris. Using a tape measure with its zero point at plot center to establish coarse woody debris transects is efficient, and it is useful later for locating the boundaries of tree plots and locating shrub and litter plots. When establishing transects for coarse woody debris, crews should mark the boundaries of subplots more than about 3 meters from the plot center. They can measure shrubs and tree seedlings immediately after woody debris, while the location of the debris transects are still fresh in their mind.

Soil Organic Layer

The default protocol assumes that there is no layer of decomposed organic material below the litter and above mineral soil. However, some sites have such a layer, and the protocol should include a separate measurement of it because such layers can contain huge amounts of carbon.

Sites that are often disturbed by plowing, fire, or other agents do not have an organic layer because it forms slowly and can be destroyed quickly. Sites with fast decomposition rates usually do not form an organic layer. Soil horizons of decomposed organic material generally occur in forest and wetland areas, when cold temperatures, moisture saturation, low pH, and nutrient limits slow decomposition and allow an input of plant material. Special cases are peat and muck soils. Peat is a buildup of minimally decomposed plant material, whereas muck is black, decomposed organic material. Under certain combinations of acid conditions, low nutrient availability, high moisture, and low disturbance, plant material can accumulate as layers of peat or muck several meters thick.

If a decomposed organic layer is present on the forest floor and included in sampling, then the material should be measured separately because its carbon content is substantially different from that of undecomposed material. To measure the decomposed organic layer, field technicians should take 16 depth measurements in an evenly spaced grid pattern spanning the litter subplot. The technicians should use a trowel to cut a vertical slice through the layer and use a rule to measure the thickness in centimeters. Quantifiers then search the literature for a density factor for the particular type of material present, as reported in fire fuels publications and ecological studies. Carbon mass is density times depth times area. (For more details, see Chapter 7.)

Appendix 9

Using Stocking Surveys to Monitor Forest Projects

Stocking surveys determine the extent to which tree species of interest cover a specified area. Such surveys usually aim to measure the survival of planted seedlings, but they may also aim to measure natural regeneration or even coverage of mature trees. Measurements of forest carbon stocks often occur only every 5 or 10 years. However, developers often need to determine if the project is proceeding according to plan—that is, that the area is substantially covered by healthy, growing trees—during the intervening years. Stocking surveys can provide a quantitative measurement of the number and distribution of trees. Surveys measuring the number and distribution of newly planted seedlings are especially important. For well-established trees, visual surveys may be more cost-effective, with a stocking survey performed if the qualitative survey indicates that stocking may be inadequate.

The target stocking density is a function of planned stand development. The acceptable range is based on knowledge of how individual trees and stands of trees develop on similar sites, the expected rate of tree mortality, and the desired level of competition among trees. Mortality does not occur linearly with age or tree size, and expert judgment is required to assess site conditions, risks, and goals, as well as to determine when a stand is considered adequately stocked and annual surveys can end.

Stocking surveys should provide two kinds of information: an estimate of the total number of trees and the proportion of growing space that they occupy. Such surveys can also provide the number of trees in each species.

The basic approach is to count the number of healthy trees on several dozen small plots. Plots should be small for three reasons. First, it is easy to miss individual seedlings or to double count seedlings on plots with a radius greater than about 4 meters. On a circular plot with a flag placed at the plot center, a field technician can walk in a circle halfway between the plot center and the plot edge, moving his or her focus from the plot center to the plot boundary and back. Technicians can easily miss small seedlings if looking more than 2 meters to each side. Sites with dense shrub or grass cover will require a more thorough search.

Second, when determining the proportion of growing space that is occupied, plots should be about the size of the tree of interest. Then the proportion of plots occupied by trees or seedlings indicates the proportion of the total growing space that is stocked. For example, suppose that the target stocking for a site is 500 trees per hectare. If plots are 1/500 hectare each, the proportion of plots occupied by at least one tree gives a good indication of the proportion of the available growing space that is occupied. A 1/500-hectare circular plot has a radius of 2.52 meters.

Third, small plots keep down the cost per plot, enabling developers to measure many for a modest total cost. Plots take only a minute or two to measure, and one person should be able to measure at least a few hundred plots well distributed over several hundred hectares in one day. Unless trees are at a very low density relative to the chosen plot size, or in a spatially clumpy distribution, 100 plots should give a 90 percent confidence interval that is not more than 10 percent of the mean estimated tree density.

Plots for stocking surveys are usually temporary, and they must be widely distributed across the area and located randomly. It is efficient to randomly start sampling along a line and then to sample at a standard interval along that line. The line can begin at the point of access to the area being sampled. It may be efficient to sample along several parallel lines or in some other pattern to cover the area. Lines may follow compass bearings, or in open lands, crews may locate them by traveling straight toward a distant landmark.

Dividing the total area to be sampled by the target number of plots gives the area represented by each plot. This guides the choice of distance between plots. Find the area represented in each plot in meters squared, and find the square root of this number. If transects are equally spaced across the entire project area, and the plots are equally spaced on each transect, then the plots would be the calculated distance apart. For example, suppose that 100 plots are to be installed in an even grid across a 100-hectare project area. One hundred divided by 100 is one, so each plot represents an area of 1 hectare. One hectare is an area of 10,000 square meters. The square root of 10,000 is 100, so plots would be 100 meters apart. In practice, transects that are not evenly distributed can still provide good estimates. They should, however, be far enough apart so that the trees on one plot have only a negligible impact on trees on any other plot. Plots should be at least 30 meters apart unless the goal is an unusually high sampling intensity.

If the goal is to measure stocking of interplantings in an area partially covered by established trees, plots should be outside the areas covered by existing trees. If the canopy of an existing tree covers any part of a plot, crews should move the plot center away from the tree until the entire plot is in the open. If existing trees are too dense for the plot to be in the open, crews should continue along the transect until they reach an open site. (This approach works only for determining whether gaps in existing forests are being filled, not for measuring total tree cover.) Do not move plots because they fall on brush clumps, debris piles, rock outcrops, roads, streams, or any other non-tree cover. The goal is to measure what exists, not an idealized version.

After measuring a plot, return to the plot center, relocate the bearing, and walk the specified number of paces to the next plot center. If check cruising will occur, flags can be left and the bearings and starting points recorded. Otherwise, an entirely new stocking survey can be done to determine whether the mean estimated result of the two surveys is similar.

Except in very steep areas, technicians do not have to adjust the distance between plots to correct for slope. If the area is very steep, locating transects so they generally traverse slopes can minimize climbing. However, care must be taken to sample ridgetops, midslopes, and valley bottoms roughly in proportion to the proportion of the site they occupy. If the slope of the transect is generally less than 50 percent (even if the terrain is steeper), and if transects are run until they reach the far side of the area rather than a specified number of paces, not correcting for slope will not substantially clump the plots.

Field crews can quickly determine whether a seedling is inside or outside a plot using a tape of the exact length of the plot radius, and attaching one end of the tape to a stake. Place the stake at the plot center, and stretch out the tape to determine the location of the plot boundary. Distances should be slope-corrected for slopes greater than 10 percent. Crews can then tally the number of healthy trees on the plot.

The protocol should specify the approach to locating transects and determining the distance along transects between plots, the minimum number of plots to be measured, and the species or species groups to be tallied separately. The protocol should also specify if plot centers are to be flagged for check cruising, along with how to document the start of each transect.

In additon, the protocol should define what qualifies as a healthy tree. Healthy seedlings usually have complete, green-colored foliage (not chlorotic), no sign of disease, unbroken stems, and only minor herbivory damage. In healthy seedlings, the ratio of leader growth to lateral growth is greater than 1. Healthy seedlings may be free to grow or overtopped—meaning nearby vegetation substantially blocks direct sun from reaching the upper part of the seedling. A seedling is also considered overtopped if nearby vegetation is growing much more rapidly and is likely to overtop it in the future. A seedling can be classed as "free to grow" if it is healthy, is not overtopped, and has survived a minimum of three growing seasons. Individual seedlings must also have good form and a high probability of remaining or becoming vigorous, healthy, and dominant over undesired competing vegetation. If browse damage is a risk, quantifiers may require that trees attain a minimum height, such as 2 meters, before being classed as free to grow.

The analysis of field data is relatively straightforward. The data are entered in a spreadsheet program, with separate rows for each plot, separate columns for each species or species group tallied, and a column for the total number of seedlings on each plot. If the goal is to find the proportion of growing space occupied, count the number of plots with at least one seedling present. If the goal is to find the density of seedlings of each species, find the mean density per plot. Common spreadsheet programs can calculate confidence intervals. Then scale up to the unit of interest, usually hectares or acres, by multiplying by the number of plots. For example, if plots average 1.5 trees, plots are 1/500 hectare each, and the goal is to find the number of trees per hectare, multiply the number of trees per plot by 500. In this case, the mean estimated number of trees per hectare is 750. Depending on the project, the users of survey data may be more interested in the mean

estimated stocking or the lower bound of a confidence interval.

Quantifiers can estimate the mortality of plantings by subtracting the number of surviving live seedlings from the number planted. Unless trees are planted at evenly spaced intervals, counting dead or missing seedlings is not a reliable way to estimate mortality.

Stocking surveys are an efficient way of assessing the effectiveness of efforts to establish trees. If a survey indicates that stocking is somewhat lower than the target rate, it is best to either measure more plots or perform another survey to determine whether the first set of plots happened to land in spots with unusually low stocking.

If the goal is to assess the stocking of larger trees, aerial photographs or satellite imagery may be useful sources of information. Quantifiers can assess stocking by overlaying the photograph with a relatively fine grid and counting the proportion of grid cells that are more than half covered by tree canopy. (Avery and Burkhart 1994 describe this method.) Quantifiers can use the same method to manually analyze remotely sensed imagery with pixels smaller than the width of the tree canopy, such as images with a 1-meter or 2-meter resolution. Quantifiers can use various algorithms to analyze satellite imagery with pixels somewhat larger than tree canopies, such as images with a 10-meter to 30-meter resolution. Satellite imagery with coarser resolution provides less reliable information on the proportion of tree cover.

Appendix 10

Determining the Density of Woody Materials

Some of the biomass equations used to convert physical and chemical measurements of forest specimens to total carbon content require knowledge of the *density* of the specimens. (Density is defined as the mass of an object divided by its volume, and typically is in grams per cubic centimeter.) When quantifying forest carbon sequestration, analysts need density if equations return volume instead of mass. Published literature can often provide the density of a particular material. If that information is not available, analysts may have to measure the density. This appendix describes how to do so.

Analysts find the density by sampling specimens, measuring the dry weight and volume of each, and dividing the mass (or weight) of the sample by its volume. Biomass equations often report and use density as a *specific gravity*. The specific gravity of a substance is simply density divided by the density of water. Because the density of water is 1 gram per cubic centimeter, the specific gravity of a substance is just a shorthand way of expressing the density of that substance in grams per cubic centimeter without having to write down the units. Expressing density as a specific gravity, therefore, requires calculating it in units of grams per cubic centimeter. For example, if wood is measured in pounds per cubic foot, divide by 62.43 pounds per cubic foot to get the specific gravity.

Trees

Densities of a variety of parts of trees have been published—most for stem wood, but some for stem bark. The densities of other tree components are usually not available. This section describes a method for determining tree density if one is not available. Although this method is for wood and bark, it can be applied to stems, branches, and roots. A similar method can be applied to foliage.

Although trees vary in the proportions of biomass that occur in their different components, using a density calculated from tree boles provides acceptable results. The wood of tree branches usually has a somewhat higher density than stem wood (McAlister et al. 2000). Bark density varies widely by species, but a tree's bark is usually less dense than its wood, so studies that develop tree densities should include bark. Roots and foliage are usually measured by mass, not volume, so density measurements of foliage and roots are not needed to calculate carbon stocks in these components. Analysts can extrapolate density that is correct for stem wood and bark to the entire tree.

Because density varies widely across species, it is desirable to determine the density of individual species. If several species are known to have very similar densities, then constructing an average for that group may be efficient. Information on the dry weight of a specific amount of lumber provides a good indication of the density of that wood. However, if analysts need densities for many species, and they group them because the cost of measuring each species is too high, estimates of mass may have greater uncertainty. The best approach is to use local knowledge to divide species into light, moderate, and heavy woods and calculate a density for each group. The USDA's *Agricultural Handbook* (AH188), available from the Forest Products Laboratory (http://www.fpl.fs.fed.us), provides the densities of several dozen U.S. and non-U.S. species. The Intergovernmental Panel on Climate Change provides the densities of several hundred tropical tree species (IPCC 2003). Unless otherwise stated, analysts should assume that densities are for stem wood only, not the whole tree.

Analysts may use a variety of methods to sample trees and calculate density. Whatever the method, analysts should weight measurements from different tree components in proportion to their relative masses. To perform such proportional weighting, do the following:

1. Make all samples the same shape (such as a complete disk 2 centimeters thick), and take samples at uniform increments along the entire length of the tree and components (such as every 3 meters along the trunk and branches).

2. Take the number of samples in each tree component in proportion to the proportion of total volume in that component.

3. During data analysis, weight the observations from samples in different components in proportion to the proportion of total volume in that component.

The approach of making all the samples the same shape and sampling at a fixed increment requires the least amount of information about the shape and mass of trees. In theory, it avoids introducing errors into the proportions of mass in different tree components. The method also systematically samples possible variations in density in each tree component. This approach is efficient for sampling the bole, but sampling branches requires a lot of effort because it results in numerous samples from small branches. Unless the sampling increment matches the length of logs, this method destroys the timber value of sampled trees.

When the number of samples taken from each tree component is proportional to the proportion of mass in that component, the density of each sample is counted equally in calculating species density. In theory, this

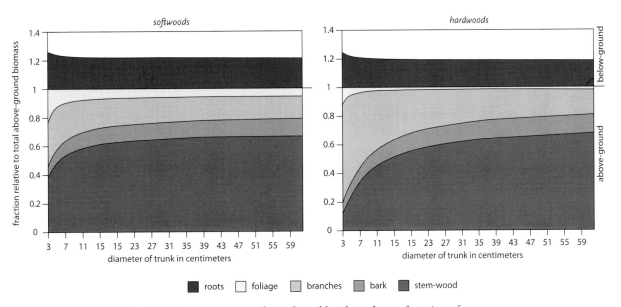

Figure A.8 Generalized fractions of biomass in softwoods and hardwoods as a function of tree diameter relative to the total above-ground biomass. Because each fraction is relative to the total above-ground biomass, the sum of the fractions, which include above- and below-ground biomass, is greater than 1.
Source: From equations in Jenkins et al. 2003.

introduces errors when tree parts of different diameters have different densities, but in practice, this error is small. This method requires a sampling strategy for proportional sampling, but limits opportunities for technicians to introduce bias by selecting nonrepresentative samples. Figure A.8 shows the proportion of mass in different components of hardwoods and softwoods as a function of diameter. When selecting sampling intensities of different tree components, analysts should use proportions of the moderate to large trees that the project expects to grow because most mass will be in these trees.

Weighing is easiest if the number of samples from each tree component is proportional to the proportion of total mass in that component. This requires estimating the proportion of mass in each component. Then, during data analysis, analysts weight the density calculated for each sample in proportion to the number of samples in its component and the proportion of total mass in the component. This method is described next. This discussion assumes that a single species is being measured. If specific gravity is being developed for multiple species, sample each species in proportion to its estimated proportion in the total mass expected to occur within the project area and increase the number of samples to account for greater variation from sample to sample.

Begin by selecting at least 10 trees spanning the range of diameters and growth forms to which the density will be applied. Analysts should use at least 30 samples, more if they are calculating one density factor to apply to multiple species. Careful analysis of 30 samples should yield a calculation of density with 95 percent confidence that the observed mean specific gravity is within 3 percent of the actual density. More trees should be sampled if the crown closure ratio varies, if the samples include both suppressed and dominant trees, or if a narrower confidence interval is desired. If densities of individual tree components are needed, each component should be sampled separately, with enough intensity to yield the desired precision.

Collect samples from predetermined points along the length of the tree bole. If logs of specific length are not needed, a simple method for obtaining four samples per tree is the following: Take one sample immediately above any butt swelling. Take a second sample halfway between the first sample and the treetop. Then take a third sample halfway between the first and second samples. Finally, take the fourth sample halfway between the second sample and the treetop. If the wood is valuable in logs of specific length, take samples at the base of the tree and at the top end of each log. If trees are felled using a mechanical harvester, the first 60 centimeters or so above the cut is damaged by the gripping mechanism, so the samples should come from above or below the damaged area.

Sample branches by randomly selecting the desired number. Many random sampling systems are easy to use. One example follows. The proportion of above ground biomass that is in branches varies widely, with a midrange estimate of 20 percent. Suppose your goal is to obtain a density of aboveground wood and bark in stems and branches. Assume that branches are 20 percent of the mass, so one branch sample is desired for every four bole samples. Technicians take three bole samples in each tree sampled. In this case, they take a branch sample from the first three of every four trees sampled, resulting in three branch samples for every 12 bole samples. For each tree where a branch sample is to be taken, draw a random number with two decimal places between zero and one. Suppose you draw 0.30. After the tree is felled, find the point three-tenths of the distance from the base of the crown toward the tree tip. Select the branch nearest this point. If branches are in whorls, use the same random number used to find the whorl to select a branch on the whorl. Starting at the closest point, travel clockwise around the bole three-tenths of the way around. Select the branch closest to this point. Draw another two-digit random number. Suppose you draw 0.60. On the selected branch, find the point six-tenths of the distance from the bole toward the branch tip. Sample at this point. This method gives equal probability of sampling any point along any branch.

Take the samples by cutting a disk, a short length of log or branch about 2 to 3 centimeters long. If the disk is greater than about 8 centimeters in diameter, cut a wedge from the disk, like a slice of pie, with the point of the wedge near the pith at the center of the log and the bark on the wide end of the wedge (see Figure A.9). The exact width of the wide end of the wedge does not matter. Wider wedges are easier to cut. Cutting a wedge that is no wider than of 3 to 6 centimeters makes the samples lighter to carry and easier to process. Narrower wedges are more likely to break. Stems, branches, and roots less than about 8 centimeters in diameter are easy to sample as entire disks. Pieces about 5 to 20 centimeters in diameter are easy to sample as a half-circle. Cutting a wedge proportionally samples heartwood and sapwood and proportionally samples juvenile wood near the center of the tree and mature wood near the outside of the tree. Samples taken from hollow trees do not have a point.

It may be desirable to proportionally sample clear stem wood and knots. Field technicians can roughly estimate the proportion of knots in the tree bole. Because branches grow in size as the trunk grows in diameter, finding the proportion of the total area of the bole surface that is occupied by branches gives a reasonable approximation of the proportion of the tree volume that is knots. To obtain an estimate of the total branch area, the number of branches protruding from the stem can be counted and multiplied by the cross-sectional area of an average branch where the branch emerges from the tree trunk. Include dead branches. The total bole area can be calculated by approximating the diameter of the tree at the base times π times the total tree height times one-half. All terms should be in the same units. Dividing the total branch area by the total bole area gives the estimate of the proportion of the tree that is knots. If the proportion of knots is 2 percent and 30 samples are being collected, then knots should compose the equivalent of about two-thirds of an average-size sample.

If the number of samples taken from each component is proportional to the mass in that component,

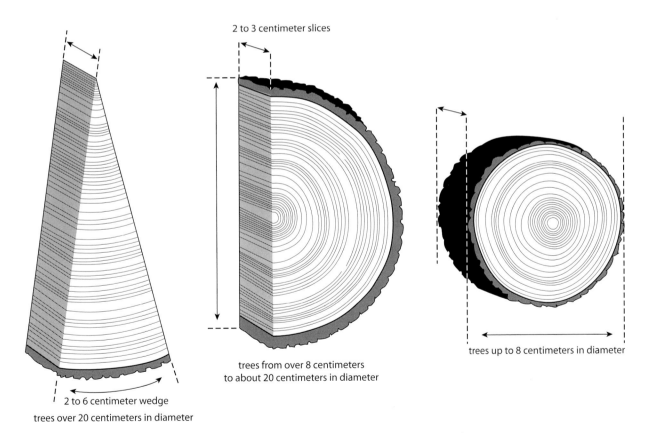

2 to 3 centimeter slices

2 to 6 centimeter wedge
trees over 20 centimeters in diameter

trees from over 8 centimeters
to about 20 centimeters in diameter

trees up to 8 centimeters in diameter

Figure A.9 Samples for measuring tree density. Wedges are generally most appropriate for pieces more than 20 centimeters in diameter. Disks are suitable for pieces up to about 8 centimeters in diameter. Half-disks are suitable for in-between sizes.

field technicians only need to record the sample number and field wet weight—theoretically. However, to help sort out anomalous numbers or to allow analysis of variation in density by tree component, size, or other variables, technicians should record additional data, such as tree number, species, diameter at breast height, tree component, date of sampling, and general location. A few words describing the stand structure and composition may be useful in deciding the kinds of stands to which to extrapolate the specific gravity. Each sample should be weighed immediately after cutting, labeled with the tree number and sample number, and sealed in a plastic bag. The sample number should indicate where in the tree the sample was taken. For example, when samples are taken from the cutting allowance of logs cut to 10-meter lengths, one might be labeled "8, 2 of 3, stem." This would indicate that the sample is from tree number 8, that it is the second sample from the base of a tree where three samples were taken, and that this sample is from the stem. Because of the pattern for locating samples, we know that this sample was taken at a little more than 10 meters plus stump height above the ground. Technicians should label the samples with a permanent marker. If a sample breaks into pieces, each piece should be labeled with the code for that particular sample.

Samples should be processed within a few days or frozen until processing to limit decay. Find the volume of each piece by submerging it in a graduated cylinder partially filled with a known volume of water and reading the volume when the sample is completely submerged. Starting with a water volume that is a round number makes it easier to inspect the calculated volumes of samples and to check to see whether the sample volume was calculated correctly. Samples generally float, but a piece of wire can be used to push them down until they are submerged. The displacement from the sample should be read quickly to limit changes that occur when water soaks into the sample. The volume of the sample is the volume of water in the graduated cylinder while the piece is submerged minus the volume of water in the cylinder before the sample was inserted. If this method is being used to calculate separate densities for wood and bark, the bark must be removed from the wood and treated as a separate sample. Some water

is lost from the cylinder with the measurement of each piece. After each measurement, inspect the water level and add more as needed.

After measuring the volume, dry the samples to find the dry weight. Oven drying is fast, but it removes volatile carbon from the sample (Lamlom and Savidge 2003) and reduces the observed density by up to 2 percent. Ideally, samples are dried at room temperature in a vacuum desiccator with a few minutes of vacuum each day. This equipment may not be available, however, and samples of the size recommended here would probably take many weeks to dry completely. Samples should be dried in an environment with low relative humidity. At 27°C and 10 percent relative humidity, the equilibrium moisture content of wood is 2.4 percent of the dry weight (Wenger 1984). If heat is used to dry the samples, the temperature should not exceed 65°C (Harmon and Lajtha 1999). Samples should be monitored and dried until weight has been stable for at least 24 hours. Record the weight of the wood component and bark component of each dry sample.

Dividing the dry weight by the green volume yields the density. Woody material shrinks as it dries, so the volume of dry samples is smaller than the volume of those samples when wet in the field. Specific gravity is applied to calculate the dry mass of trees measured in the field, and those trees are measured at wet field size. Thus the wet volume—not the dry volume—is used to calculate specific gravity.

If the number of samples from each tree component is roughly proportional to the proportion of total mass in that component, then simply averaging the densities of all samples yields the whole-tree density. The standard error can be calculated using the equation for simple random sampling.

If the mass of samples taken from each tree component is proportional to the relative masses of the different components, whole-tree density is calculated by first determining the average specific gravity of each component and then finding the weighted average of the specific gravities of each component. The weighted average is found by multiplying each component density by the proportion of the total mass in that component and then summing. Measuring components separately is like stratified sampling; methods in Appendix 2 for calculating the confidence intervals of stratified sampling can be used to find the confidence interval in this case.

If neither the mass of samples nor the number of samples from different tree components is proportional to the mass in each component, the calculation is like that for stratified sampling, for which sampling is performed at different intensities within each stratum.

Roots

In the past, it has been acceptable to calculate root mass as a function of aboveground tree size and mass. In the future, it may be necessary to measure root mass. This section describes methods for developing project-specific equations to predict root mass from aboveground tree measurements.

When root mass is calculated, often only the coarse root mass is estimated. Fine roots are usually measured with soil or their mass is ignored. For forests with large trees, fine root biomass has been calculated as less than 2 percent of the total tree and shrub biomass (Santantonio et al. 1977). Articles on the biomass of woody roots often designate the boundary between fine and coarse roots as 1-centimeter in diameter.

Technicians can sample root mass by digging up roots, washing them to remove soil, weighing them in the field, taking subsamples, drying those, and converting field weight to dry weight. Complete excavation starts at the stump and follows roots until they terminate or to a fixed distance from the center of the tree trunk. Needless to say, the process is laborious. Some studies limit the amount of work by excavating a circular trench around the tree at a fixed distance from the tree center. The area encircled by the trench often has a radius of 2 meters or less, sometimes as little as 30 centimeters. To capture the bulk of root biomass but limit the amount of work, the excavated area should have a radius that is six times the radius of the tree. On the face of the trench toward the tree, count the cut roots and measure the diameter of each. Use these measurements to estimate the mass of roots removed during trenching and present outside the trench. Excavate the area encircled by the trench and weigh the roots.

Bledsoe and colleagues (1999) recommend sampling root biomass by combining the excavation of the tree root ball and pits used to sample lateral root biomass. Pits are randomly located in areas not included in the root ball excavation. (For more information, see Bledsoe et al. 1999; Böhm 1979; and Vogt and Persson 1991.)

Woody Debris

Factors for the density and carbon content of woody debris are needed to convert field measurements of debris volume to mass. Woody debris has the added complexity that its density changes as it decomposes. Density usually decreases as wood decomposes. Adding even greater complexity, some species lose a lot of density at moderate levels of decay, whereas others do not lose a large proportion of their density until far along in the decomposition process. The density of species with decay-resistant heartwood and readily decomposed sapwood may fall as the sapwood decomposes and then rise again as the sapwood disappears and leaves relatively undecomposed heartwood.

If a project analyst must determine the density of woody debris, he or she should do so outside the plots used to measure the mass of woody debris. This is because sampling removes some wood, and cutting speeds the decomposition of the remaining pieces. Analysts can determine a set of woody debris densities for the project and then apply them to pieces measured within the sampling plots. Densities may be used for both downed wood and snags.

Sampling for density requires making several choices concerning species, decay classes, and the minimum size of pieces. Different species have different starting densities and decay trajectories. Calculating the specific gravity of individual species is more accurate, but if a project encompasses many species, doing so may be too costly. In addition, once material is fairly well decomposed, it becomes difficult or impossible to identify species in the field. Analysts can develop separate densities for undecomposed material from different species while also specifying a single density for largely decomposed material.

If grouping species, analysts should first do so according to the densities of the live wood. Common temperate species have densities from 0.31 to 0.64, and tropical species have densities from 0.24 to greater than 1.0. In a temperate region, a high-density species can have more than twice the mass per unit volume than a low-density species, and in the tropics, this ratio is greater than four. If analysts use a second variable to group species, they should try to separate species whose interior pieces remain minimally decomposed from species that decompose more evenly. For example, the core of western red cedar logs can remain sound even as the diameter of the piece shrinks, whereas true firs tend to decompose more evenly.

A five-category system of decay classes is recommended (see Figure A.10 and Table A.14):

1. Hard; bark mostly intact; branches not rotted.
2. Hard; losing bark; fine branches rotted off.
3. Soft exterior, hard interior; not totally conforming to ground topography.
4. Soft; conforming to ground topography; partially buried.
5. Decayed to chunks or mush; substantially buried.

In this system of decay classes, some species lose very little density from class 1 to class 2, whereas others lose more than 25 percent. Less decay-resistant species usually lose more density for a given visual condition. In decay-resistant species, especially those that develop large pieces of debris, class 2 density is often fairly close to class 1. The density of class 3 material is highly variable. In species that have either lost a lot of density by class 2 or are highly decay resistant, class 3 density is not much lower than class 2. The specific gravities of class 4 and class 5 material are similar for many species. Compared with the variation between species of class 1 densities, the variation in class 5 densities is modest. Most forested areas include only modest amounts of class 5 material, so class 4 and class 5 material may be sampled together to limit costs.

Harmon and Sexton (1996) provide densities by decay class for selected tropical Mexican and western U.S. conifer species. Adams and Owens (2001) provide densities for 21 Appalachian hardwood species using three categories of decay. Their class 1 is comparable to class 1 presented here. Their class 2 is equivalent to classes 2 and 3 here. Their class 3 is equivalent to classes 4 and 5 here.

Sometimes only two decay classes are used: hard and soft. This approach is best for systems with very fast decomposition rates and low levels of woody debris. Field technicians can distinguish the classes by firmly striking a piece of detritus with a machete. If the strike yields a firm thunk or a hollow or ringing sound, the piece is classified as hard. If the sound of the strike is muffled, soft, or squishy, the piece is classifid as soft.

Fine woody detritus is often defined as 1 to 10 centimeters in average diameter. Coarse debris is often defined as greater than 10 centimeters in average diameter. Harmon and Sexton (1996) recommend designating coarse debris as greater than 10 centimeters in diameter and more than 1.5 meters in length.

Once analysts have chosen the categories to measure, field crews should collect samples within each category. More samples are needed to get a tight estimate of specific gravity for debris than for live wood because the

Table A.14 Average Density of Woody Debris by Decay Class

Species group	Class 1	Class 2	Class 3	Class 4	Class 5
Tropical; high density; decay resistant	0.779	0.726	0.669	0.463	0.225
Tropical; high density; not decay resistant	0.619	0.441	0.300	0.200	0.200
Tropical; low density	0.288	0.211	0.096	0.096	0.114
Temperate; moderate density	0.404	0.327	0.293	0.172	0.158
Temperate; low density	0.329	0.296	0.225	0.170	0.143

Note: Units are grams per cubic centimeter.
Source: From Harmon and Sexton (1996).

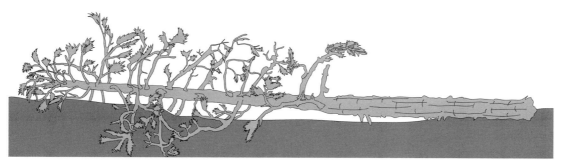

Class 1: Fine branches not decomposed, bark intact, wood not decomposed since falling.

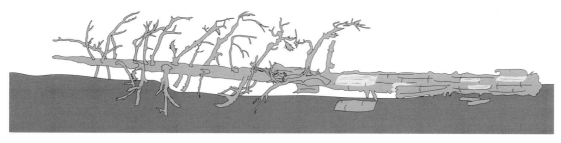

Class 2: Limbs and most bark present, fine branches decomposed, sapwood becoming soft and fibrous.

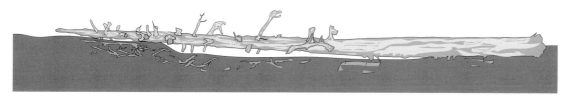

Class 3: Only limb stubs present, sapwood soft or missing and heartwood beginning to soften and become fibrous.

Class 4: Wood soft and fibrous or cubical, bole starting to conform to ground.

Class 5: Decomposed to cubical reddish brown blocks, possibly encased in a hard shell, generally substantially buried.

Figure A.10 Classes of the decay of woody detritus. The state of decomposition of woody debris is categorized by "class." The greater the class, the more decomposed the debris.
Source: Adapted from R. Bartels et al. 1985, Dead and down woody material, in *Management of Wildlife and Fish Habitats in Forests in Oregon and Washington*, ed. E.R. Brown, Washington: U.S. Government Printing Office, no. R6-F&WL-192-1985, 171–86.

density varies within each debris class. Thirty samples per decay class are likely to yield an estimate with a 95 percent confidence interval that is about 5 percent of the estimated mean specific gravity. Ten samples per class generally yield an estimate with a 95 percent confidence interval that is about 10 percent of the estimated mean. To limit costs, class 4 and class 5 can be lumped together and only ten samples collected from class 1 material. Lumping class 4 and class 5 increases the estimation error only slightly because class 4 and class 5 densities usually differ minimally and most sites have little class 5 material. The sample size of class 1 material can be reduced because density is usually less variable within that class.

Decomposition often varies along the length of a piece of detritus, especially in the middle stages of decay. To minimize sampling bias, samples should be taken at a random point along the length of a piece of debris. When technicians encounter a piece they will sample, they can draw a random number between zero and one. Starting at the larger-diameter end of the piece or the first end encountered if no end is obviously larger, they can move along the length of the piece the proportion of the total distance matching the random number. For example, if the random number is 0.4, they move 40 percent of the length of the piece, starting from the large end, and sample at that point.

If the piece is solid enough to saw, they can cut a slice of material perpendicular to the long axis of the debris. This is like sampling to determine the density of wood. Because wood in contact with soil decays differently from wood exposed to the air, reducing the size of individual samples by randomly cutting a wedge from a complete disk is not advisable. Instead, cut a wedge from the part of the disk that was against the ground, and cut a second wedge from the part of the disk that was away from the ground (see Figure A.11). A sharp chainsaw will cut even relatively crumbly material if the chain is brought to full speed before it contacts the debris. If the material is prone to crumbling, technicians should record the dimensions of each piece immediately after cutting it and before bagging and transporting it so analysts can calculate the volume later from the dimensions. This method may require cutting the sample free from the rest of the detritus and gen-

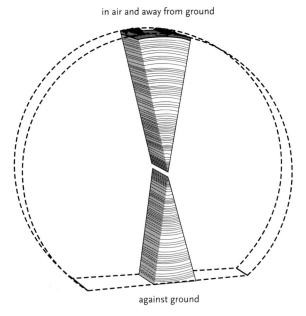

in air and away from ground

against ground

Figure A.11 Sampling large pieces of partially decomposed woody debris. Because wood in contact with soil decays differently from wood exposed to the air, reducing the size of individual samples by randomly cutting a wedge from a complete disk is not advisable. Instead, cut a wedge from the part of the disk that was against the ground, and cut a second wedge from the part of the disk that was away from the ground. The density of the sample is found by dividing mass by volume. The specific gravity of a class of debris is found by averaging all samples of that class. Confidence intervals can be calculated for each class using the equation for a simple random sample.

tly clearing the detritus away from the sample without moving the sample. Material that crumbles into chunks or mush can be scooped into a bag, with mineral soil carefully excluded.

Using methods similar to those for measuring the specific gravity of wood, samples should be labeled, their volume should be measured, then samples should be dried, and their dry weight should be measured. This process assumes that all samples are dried to obtain dry weight, so there is no need to measure field wet weight. When measuring the volume of decayed fragments, analysts may need to use a screen or other device to push the fragments below the surface of the water. Analysts may have to dump the entire contents of the graduated cylinder used to measure volume into a relatively fine mesh sieve to collect all the fragments.

Litter

Quantifiers can find litter density by recording the area of the ground from which the litter was collected, measuring the thickness of the litter layer, and weighing the sampled litter. Because litter depth may vary significantly between points a few centimeters apart, field crews should take several depth measurements on each subplot. (Sixteen measurements in a 4-by-4-meter grid pattern can provide a reasonably accurate view of average litter thickness.) Multiplying litter depth by surface area yields the volume of litter collected, and dividing this volume into the mass of litter collected yields the density. Litter density often changes with the time of year, so quantifiers should use density factors that are appropriate to the time of sampling. One way to do this is to calculate the litter density for the project from measurements made on the first 20 or 30 plots sampled. Litter density can also vary across vegetation types. (If sampling is stratified to reflect different types of vegetation, quantifiers may develop a litter density for each stratum or an average density for all strata.)

Processing Litter Subsamples

When measuring the proportion of carbon in litter, quantifiers can use subsamples collected for determining the ratio of dry weight to wet weight for carbon analysis. If biomass sampling is stratified, they should consider whether to use the same stratification for analyzing litter carbon. Stratification of biomass sampling is usually driven by variation in productivity or management, whereas variation in litter carbon stocks usually depends more on decomposition rates.

After drying and weighing litter subsamples to determine dry weight, quantifiers can use them to find the proportion of carbon in litter. The reported variation in the proportions of carbon in different plant components suggests that 10 samples will probably give a confidence interval of about 2 percent. Because samples are collected for other reasons, the cost of using them to determine carbon content is the cost of grinding and running them through an analysis. This cost should be modest, and quantifiers should analyze at least 30 to 40 samples to measure the carbon content of litter.

To measure the carbon content of wood, litter, or detritus, quantifers use infrared absorption or gas chromatography to quantify the carbon dioxide emissions from dry combustion of the subsamples.

Appendix 11

Correcting for Degree of Slope

Carbon stocks are calculated per unit of area, with the area measured on a flat, horizontal plane. If the land is flat and smooth, the actual surface area of soil equals the horizontal area. However, land is often not flat. As land becomes steeper and more dissected, the surface area becomes ever greater than the horizontal area (Figure A.12). Quantifiers and crews need to correct for this effect when determining plot size because errors multiply as quantifiers expand carbon stocks on plots to estimate stocks across an entire project.

There are two ways to correct for land slope. The recommended method is to do so in the field, using *slope distance*, which reflects both horizontal distance and slope. The other approach is to use plots of uniform size to make field measurements and then use slope distance to mathematically expand the measured area during analysis. The latter approach slightly reduces the sampling intensity but requires recording the slope observed on each plot. On large plots that include changes in terrain, assigning a single slope is difficult. If slope correction occurs during data analysis, the correction expands an ellipse to a circle as a function of slope (the equation is not presented here).

On slight slopes, the slope distance is almost the same as the horizontal distance. Protocols should specify a slope angle above which slope distance is corrected, usually 10 to 15 percent. On slopes flatter than the threshold angle, crews can ignore slope correction. On a 10 percent slope, the correction is to add 0.5 percent to the distance being measured. On a 15 percent

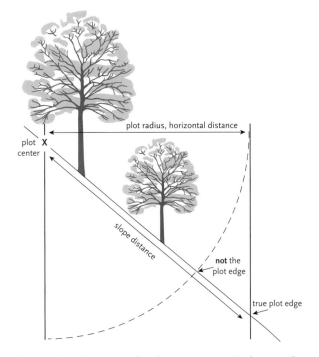

Figure A.12 Correcting for sloping terrain. Carbon stocks are calculated per unit of area, with the area measured on a flat, horizontal plane. If the land is flat and smooth, the actual surface area of soil equals the horizontal area. However, land is often not flat. As land becomes steeper and more dissected, the surface area becomes ever greater than the horizontal area. Quantifiers and crews need to correct for this effect when determining carbon stocks. The actual horizontal distance is obtained from measurements of the slope distance and plot slope.

Table A.15 Slope Correction

Slope (percent)	Slope distance (meters)				
	0.1-hectare plot	0.01-hectare plot	0.002-hectare plot	0.005-hectare plot	0.001-hectare plot
0	17.84	5.64	2.52	3.99	1.78
5	17.86	5.65	2.53	3.99	1.79
10	17.93	5.67	2.54	4.01	1.79
15	18.04	5.71	2.55	4.03	1.80
20	18.19	5.75	2.57	4.07	1.82
25	18.39	5.82	2.60	4.11	1.84
30	18.63	5.89	2.63	4.17	1.86
35	18.90	5.98	2.67	4.23	1.89
40	19.22	6.08	2.72	4.30	1.92
45	19.56	6.19	2.77	4.37	1.96
50	19.95	6.31	2.82	4.46	1.99
55	20.36	6.44	2.88	4.55	2.04
60	20.81	6.58	2.94	4.65	2.08
65	21.28	6.73	3.01	4.76	2.13
70	21.78	6.89	3.08	4.87	2.18
75	22.30	7.05	3.15	4.99	2.23
80	22.85	7.23	3.23	5.11	2.28
85	23.42	7.40	3.31	5.24	2.34
90	24.00	7.59	3.39	5.37	2.40
95	24.61	7.78	3.48	5.50	2.46
100	25.23	7.98	3.57	5.64	2.52

Notes: This table shows slope distances (in meters) by slope (in percent) for circular plots of given areas. To use this table, measure the slope and find the column for the plot size. The number in this cell is the slope distance, which gives an accurate horizontal distance for the plot. For example, if the slope angle in the direction for which the plot edge is being determined is 30 percent, and crews are trying to find the edge of a 0.01-hectare circular plot, they would measure 5.89 meters along the ground surface. Crews measure the slope angle along the line between the plot center and the point on the edge of the plot they are trying to determine. The slope angle will vary as the direction from plot center varies. For example, when measuring straight across a large, smooth hill slope, crews will use a slope of zero, even though the slope might be quite steep if they travel straight uphill or downhill from plot center.

slope, the correction is to add 1 percent to the distance being measured. The threshold for slope correction should be set at about the limit of the precision expected of field crews. If crews are expected to measure distances to within ± 0.5 percent, then the slope above which distances should be corrected is 10 percent.

For modest slopes, when crews are measuring distances with a tape, they can often correct for slope by simply holding the measuring tape horizontally. Similarly, when they are measuring distances using an electronic device, they can often measure horizontally. When using a rigid frame to determine subplot boundaries, such as a 0.5-meter-by-0.5-meter square for sampling litter, technicians can correct for slope by holding the frame horizontally, resting it on the ground at the high point of the subplot boundary. Crews can then cut the litter along the line directly below the frame edge.

On steep slopes, and sometimes when measuring uphill, technicians will not be able to measure horizontally. In those cases, they should measure the slope angle and use a slope correction table to determine the appropriate distance for locating the plot boundary. Field crews must have slope correction tables for each distance they will be measuring. (See Table A.15 for distances used in the default protocol.)

Although this manual uses percent as the units for measuring slope, quantifiers could use degrees instead. Whatever the unit, they should use it exclusively, as a correction table for one unit is not accurate for the other. Furthermore, conversion between percent and degrees is not linear. A zero-degree slope is the same as a zero-percent slope, but a 45-degree slope is a 100 percent slope. A 90-degree slope is a perfectly vertical cliff, which has an infinite-percent slope.

Slope correction tables can be created for any distance. When slopes are measured in percent, an equation for calculating slope distance is based on the Pythagorean theorem, which states that for a right triangle, the squared length of the hypotenuse equals the sum of the squared lengths of the other two sides. The slope distance is the length of the hypotenuse, and solving for that yields the equation

$$D_s = \sqrt{\left(D \times \frac{s}{100}\right)^2 + D^2} \qquad \text{Equation A11.1}$$

where D is the desired horizontal distance, D_s is the distance to be measured along the slope, s is the slope angle in percent.

If slope is measured in degrees, crews can calculate slope corrections in the field using a calculator with trigonometric functions. However, this creates an opportunity for introducing mistakes in key punching, and it is safer to use a slope correction table even when working in degrees. When calculating the slope correction with slope in degrees, crews need few keystrokes to use the equation

$$D_s = \frac{D}{\cos S} \qquad \text{Equation A11.2}$$

where D is the desired horizontal distance, D_s is the distance to be measured along the slope, S is the slope angle in degrees, and cos is the cosign trigonometric function.

Note that some spreadsheets and calculators require users to enter the angle in radians instead of degrees. If this is the case, crews must convert degrees to radians. Cosign is less than one (except it is equal to one where slope is zero), so the slope distance will always be greater than or equal to the horizontal distance.

Calculating Carbon Stock and Changes in Carbon Stock

This appendix describes how to calculate carbon stock and changes in carbon stock from data collected using the default method discussed in Chapter 6. This appendix begins by describing calculations made using measurements from each subplot. Subplot measurements are scaled to common units of mass and area, such as tons per hectare, and summed.

Litter

Field crews measure the mass of litter within a square frame that is 0.5 meters on each side. If the frame is horizontal, it covers an area of 0.25 square meters. A subsample of litter is collected, weighed in the field, taken to a lab, dried, and weighed again. The ratio of dry to wet weight is found by dividing the dry weight of the subsample by its field wet weight. The dry weight of the litter sample is found by multiplying the field wet weight of the entire litter sample from the 0.25-square-meter plot by the dry-to-wet ratio for that plot. A hectare is 10,000 square meters, so the plot is 1/40,000 hectare. The litter mass on a per-hectare basis is found by multiplying the sample mass by 40,000. Quantifiers convert grams to tons by dividing by 1,000,000 to yield the litter mass in metric tons per hectare.

If the frame was not horizontal when the sample was collected, slope correction is required. On a slope, the frame can be held horizontally and the litter cut along the line directly below the inside edge of the frame. If crews do not use this procedure, and instead lay the frame directly on a slope, the area sampled is less than 0.25 square meters. Slope correction expands the area sampled to 0.25 meters (see Appendix 11).

When slopes are recorded, recording them in percent makes calculations less complex. When slopes are measured in percent, the slope correction equation is

$$W_C = W_D \times \sqrt{1 + \left(\frac{s}{100}\right)^2} \qquad \text{Equation A12.1}$$

where W_C is the slope-corrected dry weight of the litter on the subplot, W_D is the measured dry weight of the litter on the subplot, and s is the slope angle of the subplot measured in percent.

Note that when the slope is zero, the corrected weight is the same as the uncorrected weight. For positive slopes, the corrected weight is always greater than the uncorrected weight because a frame covers less horizontal distance when it is tipped than when it is horizontal.

For example, say the measured dry weight of litter collected with the frame placed on the slope is 1,500 grams. If the slope is 20 percent, then

$$W_C = 1500 \times \sqrt{1 + \left(\frac{20}{100}\right)^2} = 1500 \times \sqrt{1 + 0.04} \approx 1530$$

If the slope is measured in degrees, the equation to correct for slope is

$$W_C = W_D \times \sec s \qquad \text{Equation A12.2}$$

where W_C is the slope-corrected dry weight of the litter on the subplot, W_D is the measured dry weight of the litter on the subplot, and sec s is the secant of slope angle s where the slope angle is measured in degrees.

As with the correction when slope is measured in percent, the corrected weight is greater than the uncorrected weight because the sloping frame covers less horizontal distance than a frame in a horizontal position. Note that many trigonometric functions in electronic spreadsheets require degrees to be converted to radians.

Woody Debris

Methods for calculating the mass of woody debris are similar for fine and coarse debris. However, an adjustment factor may be added to the calculation for fine debris to correct for the fact that the long axis of small pieces may not be parallel to the ground surface. The lengths of the transects for the two types of debris are different, but the equations are the same. A shorter transect is needed for fine debris because fine pieces occur more frequently than coarse pieces.

The equation for calculating the volume of coarse debris is

$$V = \frac{\pi^2 \times \sum d^2}{8L}$$

Equation A12.3

where V is the volume in cubic meters per square meter, π is the constant 3.1415927, d is the diameter in meters of each piece of debris, and L is the length of the transect in meters (van Wagner 1968).

This equation works for transects and debris of any length, as long as all inputs and outputs are in the same units. For example, if the transect lengths and piece diameters are in feet, the volume would be in cubic feet per square foot. Quantifiers convert volume per square meter to volume per hectare by multiplying by 10,000. They convert volume of dry biomass in cubic meters to metric tons by multiplying by the specific gravity of woody debris. They convert dry biomass to carbon by multiplying by the carbon content of woody debris.

For analyzing project data, quantifiers can enter this equation into a spreadsheet program. If they do, they can enter the diameter, species, and decay class of each piece in a row. Then they calculate the volume per hectare represented by that piece, with reference to the transect length. They use conditional statements or lookup tables to select a specific gravity as a function of species and decay class. Specific gravity is used to calculate mass. Next, they calculate carbon content as a function of mass and the proportion of the mass that is carbon. Then they sum all pieces of debris on each plot to find the number of tons of carbon per hectare measured on each plot.

Equation A12.3 applies when all the factors, such as piece diameters and transect length, are in the same units, such as meters, and volume is in cubic meters per square meter. Quantifiers can modify this equation for different units. In the United States, much forestry work is performed in English units, but GHG offsets are denominated in metric tons. To find volume in cubic meters per acre, when diameters are in inches and the length of the transect is in feet, use the following equation:

$$V = \frac{10.569 \times \sum d^2}{L}$$

Equation A.12.4

where V is the volume in cubic meters per acre, d is the diameter of each piece of debris intercepted on the transect in inches, and L is the length of the transect in feet.

If quantifiers need very accurate numbers for fine debris, they should use an additional adjustment for this material. This may be needed because the transect method assumes that the long axis of each piece of debris is horizontal. The equation sums probabilities that each piece will lie flat on the ground with its long axis at a random angle to the transect and that intersections will occur at random points along the length of each piece. Large pieces of debris almost always lie relatively flat on the ground, but small pieces of debris often do not lie flat. As the long axis of a piece of debris becomes more vertical, the horizontal distance covered by its long axis decreases and the probability of intersection by a transect decreases. As a result, debris that is not flat is undercounted.

The correction factor for the slope of the long axis is the secant of that slope (Brown and Roussopoulos 1974). Harmon and Sexton (1996) present factors of 1.11

for hardwood pieces 0.63 to 2.54 centimeters in diameter, 1.03 for hardwood pieces 2.54 to 7.62 centimeters in diameter, 1.13 for conifer pieces 0.63 to 2.54 centimeters in diameter, and 1.10 for conifer pieces 2.54 to 7.62 centimeters in diameter. In practice, ignoring this correction factor means that woody debris mass is undercounted only slightly. That is because pieces smaller than 1 centimeter in diameter are included in litter, pieces larger than a couple of centimeters in diameter have a small correction factor, and very little mass in pieces would have a correction larger than 1.10 (and even then the correction factor would not exceed 1.13). Harmon and Sexton also provide factors for merely counting the number of intersections of pieces of small debris of given size classes, rather than measuring the diameter of each small piece.

If crews measure the lengths and multiple diameters of individual pieces of wood, quantifiers use different equations to calculate volume. If the quantifier would like to use a method that relies on algebra rather than trigonometric functions to expand plot measurements to volume, crews can measure the sizes of individual pieces of woody debris on plots with fixed areas. These methods are more time consuming than transect-intercept methods, but they are easier to understand, so observers may accept them more readily. Note that when crews measure woody debris on plots with fixed areas, they should measure only the parts of pieces within the plot boundary. If a piece of debris is only partly within a plot, the piece is measured as if it ends at the plot boundary. If crews measure material outside the plot boundary, quantifiers will overestimate the mass of debris.

When length and end diameters are measured for each piece of debris, the volume of each piece can be calculated as a section of a cone:

$$V = L(A_b + (A_b \times A_t)^{0.5} + A_t)/3 \qquad \text{Equation A12.5}$$

where V is the volume in cubic meters, L is the length of the piece in meters, A_b is the area of the base of the piece in meters squared, and A_t is the area of the top of the piece in meters squared (Harmon and Sexton 1996).

Recall that the area of a circle is

$$A = \pi \times r^2 \qquad \text{Equation A12.6}$$

where A is the area and r is the radius. If the piece is substantially ellipsoid, the large and small diameters should be measured and the cross sectional area calculated using the appropriate equation for an ellipsoid. The equation for the area of an ellipse is:

$$\text{Area} = \pi \times (D_{max}/2) \times (D_{min}/2) \qquad \text{Equation A.12.7}$$

where D_{max} is the length of the maximum diameter and D_{min} is the length of the minimum diameter.

Quantifiers commonly calculate the volume of pieces of woody debris as sections of a cone, an approach that is relatively accurate, especially if tree trunks are broken into sections. Trees are not actually cone shaped—trunks are usually closer to a paraboloid shape (like a cone swollen toward the tip) above any swelling at the base. Trees with butt swelling are more accurately described as a neiloid shape.

The equation for a paraboloid is

$$V = L(A_b + A_t)/2 \qquad \text{Equation A12.8}$$

where V is the volume in cubic meters, L is the length of the piece in meters, A_b is the area of the base of the piece in meters squared, and A_t is the area of the top of the piece in meters squared (Harmon and Sexton 1996).

The paraboloid equation is particularly appropriate to snags and stumps that are decomposed down to a blob of material and that have a top diameter considered to be zero (Harmon and Sexton 1996). (See the literature on tree taper equations for more detail and species-specific equations.)

Once quantifiers calculate the volume of debris, they can calculate carbon mass. Biomass is calculated by multiplying volume by specific gravity. (See Table A.14 for default specific gravities for woody debris by decay class.) Quantifiers should use any specific gravities measured on the project instead of default values. They calculate carbon mass by multiplying the total weight of the biomass by the proportion of that mass that is carbon. They can assume that the proportion of the mass that is carbon in detritus is the same as the proportion in undecomposed wood of the species (Sollins et al. 1987). (See Table A.13 for the proportions of carbon in different species of wood.)

Standing Dead Trees

The appropriate method for calculating the biomass of a snag varies by its decay class. The methods presented here start with the least decomposed and progress to the most decomposed. Decomposition classes for coarse woody debris parallel those for standing dead trees, and adjustments to biomass estimates to account for changes in density as material decomposes are similar. Stumps are treated as snags. Regardless of height, their top diameter and base diameter are measured and recorded, not their diameter at breast height.

The mass of freshly dead trees can be calculated as if the tree were alive. The foliage of many freshly dead trees is gone, so a biomass equation that includes foliage would usually overestimate the biomass of these trees by 2 to 7 percent, depending on tree species and size. However, most projects will have few freshly dead trees, so the resulting overestimation of foliage biomass will be small relative to total biomass.

If, because of some unusual mortality event, a significant proportion of the project biomass is in freshly dead trees, quantifiers can estimate the mass of foliage and subtract it from the total aboveground biomass. BIOPAK provides equations for estimating the foliage biomass of a number of North American species. If a species-specific equation is not available, quantifiers can use the equations of Jenkins et al. (2003) listed above.

Quantifiers can reasonably assume that freshly dead trees have the same specific gravity as live trees, and they do not need to adjust for loss of biomass from decomposition. When fine branches are present, quantifiers can assume that coarse roots have not yet decomposed and use the same method used to estimate the coarse root biomass of live trees to estimate the coarse root biomass of freshly dead trees. Fine root biomass is usually included in assessments of soil carbon, not tree carbon. If so, quantifiers do not need to account for rapid decomposition of fine roots after tree death.

Standing dead trees in decay class 2 are defined as having hard wood in the boles, but with fine branches rotted off and possibly missing bark. Such trees are a borderline case. Quantifiers could estimate their mass using methods for trees in decay class 1, but it is more conservative to use the methods (described below) for trees in decay class 3.

Regardless of which method quantifiers use to calculate the biomass of class 2 trees, they should account for loss of density from decomposition. Estimates of the biomass of a live tree should be reduced by the ratio of the density of the decay class and species to the density of that species when alive:

$$B_D = B_L \times \left(\frac{D_D}{D_L} \right)$$ Equation A12.9

where B_D is the biomass of the partially decomposed tree, B_L is the biomass that tree would have if alive, D_D is the density of that species of tree in the particular decomposition class, and D_L is the density of that species of tree when alive. The two density numbers must be in the unitless factor specific gravity or must be in the same units, such as pounds per cubic foot.

Quantifiers can assume that root decomposition occurs at the same rate as aboveground decomposition. With this assumption, quantifiers calculate the aboveground biomass first, adjusting for loss of mass from decomposition. They then assume the same proportion of coarse root biomass to aboveground biomass as for a live tree of that species and size and multiply this proportion by the aboveground biomass to estimate the remaining coarse root biomass.

Trees in decay class 3 are defined as having substantial rot in the bole but a generally hard core. These trees will be missing their tops as well as most branches and bark. Quantifiers should estimate the mass of these trees by calculating their volume and multiplying by the specific gravity.

The process for calculating the biomass of snags resembles that for calculating the biomass of coarse woody debris on plots with fixed areas. Quantifiers first calculate the volume of each piece, then multiply by its density to calculate mass. Unlike when calculating coarse woody debris, quantifiers do account for the remaining coarse root biomass. The most common approach to estimating the volume of a snag is to assume that it is a truncated section of a cone.

When calculating the volume of a snag as a conic

section, quantifiers calculate the volume of each tree. As discussed earlier, trees are not actually cone shaped. Above any swelling at the base, tree trunks are generally closer to a paraboloid shape. The equation for a paraboloid is: $V = L(A_b + A_t)/2$.

Volume equations that account for taper have been developed for a variety of tree species used for timber, such as pines grown in the southeastern United States. Many of these equations require only diameter and height as inputs. However, these equations are very specific and generally apply to a single species, and they may apply only to trees grown under certain stand histories. For example, thinning has been shown to make boles more parabolic in shape and increase the amount of biomass in branches and foliage (Baldwin et al. 2000). (For more information on estimating tree volume as shapes other than conic sections, see Wenger 1984 and Harmon and Sexton 1996. Pillsbury and Kirkley 1984 describes methods for measuring hardwood tree volume as a sum of stem and major branch segments.)

Once quantifiers have calculated the volume of debris, they can calculate carbon mass by multiplying volume by specific gravity. (See Table A.14 for default specific gravities for woody debris by decay class.) As described earlier, quantifiers calculate carbon mass by multiplying the total biomass weight by the proportion of that mass that is carbon. They can assume that the proportion carbon in detritus is the same as the proportion in undecomposed wood of the species (Sollins et al. 1987). (See Table A.13 for proportions of carbon in different species of wood.)

Snags in decay class 4 are defined as soft—with no remaining hard, minimally decomposed core. Snags in decay class 5 have decomposed to a mound of pieces of woody detritus. Quantifiers can calculate the volume of snags in decay classes 4 and 5 using the same method as for snags in decay class 3. If quantifiers calculate snags in decay classes 2 and 3 as sections of cones, they may want to calculate snags in decay class 4 and 5 as paraboloids. For substantially decomposed trees and blobs, the top diameter is considered zero (Harmon and Sexton 1996). Quantifiers convert volume to mass using the same method as for snags in decay class 3, but they use the specific gravity for the appropriate class.

Live Trees

Quantifiers calculate live tree mass using the methods described in the section on measuring live trees. Equations estimate aboveground biomass as a function of tree species, diameter, and height, and a different equation estimates below-ground mass. The two masses are summed to get the total biomass for the tree. Quantifiers use a species-specific carbon ratio to convert biomass to carbon.

If field crews measure larger trees than analysts used to develop biomass equations, quantifiers must use other equations to calculate the mass of those trees. Using biomass equations for inappropriately large trees will almost certainly result in substantial errors.

If trees have broken tops that are more than a few centimeters in diameter but less than one-third the diameter at breast height, and quantifiers are using an equation for intact trees (rather than a conic section) to estimate biomass, they should use a taper equation to calculate the mass of the missing top. The taper equation should be used to estimate the total tree height as if the top had not been broken. Quantifiers use the total height to calculate the mass of the total tree and then subtract the mass of the missing top. If few broken-topped trees are present, quantifiers may simply calculate the mass as if each tree were of the observed height but with no broken top. This method will underestimate the mass of the tree, but the total error in the carbon calculation should be small. If trees have a broken top with a diameter greater than about one-third the diameter at breast height, quantifiers can calculate the bole mass as a truncated conic section and multiply by wood density—similar to the method used to calculate mass of woody debris. This underestimates aboveground biomass because it does not count branch and foliage mass or the bulging shape of most tree trunks. However, the total error should be small.

Quantifiers sum the tree masses on each subplot. They then divide the mass on each subplot by the area of the subplot in hectares to yield the mass per hectare. These masses per hectare are summed across the subplots.

Calculating Total Sequestration

If the project is using stratification, the next step is to group plot measurements by stratum. Quantifiers can find the average change in carbon stocks in the stratum by averaging the amounts observed on the plots within the stratum. They should also calculate the standard deviation and standard error for each stratum. (In making these calculations, the sample size n is the number of plots in the stratum.) Quantifiers calculate the total change within the stratum by multiplying the average change in carbon stock per hectare by the number of hectares in the stratum.

Quantifiers calculate the total project sequestration by summing the sequestration on all strata. They calculate errors and uncertainties by taking a weighted average of errors on each stratum. The weighted standard error is calculated as

$$s_{\bar{y}_s}^2 = \sum_{h=1}^{k} \left(\frac{n_h}{n}\right)^2 \times s_{\bar{y}_h}^2 \qquad \text{Equation A12.10}$$

where $s_{\bar{y}_s}$ is the standard error of the estimated mean sequestration across the entire project area, across strata $h = 1$ to k; n_h is the number of plots in stratum h; n is the number of plots in all strata; and $s_{\bar{y}_h}$ is the standard error estimated for stratum h.

Appendix 13

Adapting Biomass Equations from One Species to Another

If two species have similar growth forms, a biomass equation for one species can be adapted to give reliable results for the other species. Growth forms are similar if trunk tapers at different tree diameters and heights are similar, numbers and sizes of branches are similar, and bark thickness is similar. Similar foliage density is less important because foliage composes only a small portion of tree biomass. The key consideration in adapting an equation from one species to another is the density of each species; quantifiers should make an adjustment if the densities differ. Species with similar growth forms often have significantly different densities.

To adapt the density of one species to another species, multiply the predicted mass by a density adjustment factor:

$$F = \frac{D_N}{D_O} \qquad \text{Equation A13.1}$$

where F is the density adjustment factor, D_O is the density (or specific gravity) of the original species for which the equation was created, and D_N is the density of the new species to which the equation will be applied. The two densities must be in the same units.

Analysts have developed relatively few equations for large trees, and even fewer equations use both height and diameter to predict the biomass of large trees. This is not surprising because cutting down and weighing several very large trees is onerous. Quantifiers may adapt an equation for large trees of a different species to the species of interest. The two species should have similar growth forms. Assuming that the quantifier has a valid equation for smaller trees of the species of interest, and if the diameter ranges of the two equations overlap, quantifiers can test whether the equation for big trees gives reasonable estimates of the biomass of trees of the species of interest. This testing can be done by using both equations to predict the mass of a tree of a diameter for which both equations are valid. If the two predicted masses are similar, and the two species have similar growth forms, then quantifiers can apply the equation for large trees of one species to large trees of the other species.

Appendix 14

Developing New
Biomass Equations

When appropriate biomass equations are not available, quantifiers must develop their own. This appendix describes the concepts on which biomass equations are founded, methods for sampling trees for creating biomass equations, and methods for analyzing sample data to generate biomass equations.

The usual approach to developing biomass equations is to cut down and weigh a number of trees, ranging from the smallest to the largest, to which the biomass equation will be applied. Subsamples are dried to determine the ratio of dry to field wet weight, and the dry weight of aboveground tree biomass is calculated from measured field wet weights. Regression is used to develop an equation from the measured tree weights. Root mass is predicted as a function of species and diameter. (See Bledsoe et al. 1999 for methods for sampling root biomass.) In a forest, roots intertwine and even grow together, so, other than for main roots, assigning a root to a stem may be impossible.

Choosing the Range of Trees

The first step in developing a biomass equation is choosing the tree population to which the equation will apply. The population boundaries include geographic area, type of tree (or trees), and their sizes. The geographic boundaries may coincide with project boundaries or they may be broader. The range of tree diameters must reflect the full range of diameters to which the equation will be applied. Applying equations to trees outside the sizes for which the equations were developed may not yield accurate estimates of biomass.

Tree biomass, as a function of height and diameter, varies across the geographic range of a species (Clark 1987). Because the shape of trees changes with different stages of maturity, it makes sense to separate young trees with rapid height growth from mature trees that have very little height growth. By this logic, the dividing line between small and large trees should be roughly the diameter where change in height becomes much smaller relative to change in diameter. Tree equations fit better when developed separately for small trees and large trees, with the split occurring around 40 centimeters diameter at breast height (Yuancai and Parresol 2001). Quantifiers might otherwise select a split that matches the division between classes of log use so a single equation applies to all trees in a single class (Clark, Phillips, Frederick 1986).

Although sampling large trees takes much more effort than sampling small trees, trees as large as will be observed on the project area must be sampled to develop biomass equations. It is not reliable to use equations to predict the biomass of trees larger than the trees from which the equations were developed. The absolute error in biomass predictions will be greatest in predictions for large trees, so limiting total error requires obtaining the best possible estimates of the masses of large trees.

If very large trees cannot be destructively sampled, the volume of their boles and major branches can be measured by intensive but nondestructive means, and the mass can be calculated using information on density or specific gravity from medium-sized trees (see Appendix 10). Pillsbury and Kirkley (1984) present a method for categorizing and measuring tree branch volumes. Parresol (1999) shows how to estimate the masses of small branches and use ratios. Valentine, Trotton, Furnival (1984) and Gregoire, Valentine, Furnival (1995) provide a method for subsampling and estimating branch mass.

For projects that will grow trees, developing equations for large trees that do not yet exist within the project boundary requires sampling trees outside the boundary. The climate and productivity of selected lands should match those of project lands as closely as possible.

The history of a stand affects the shape of the trees and thus the mass as a function of height and diameter. For a given height and diameter, trees that develop with wide spacing have different mass than trees that develop with tight spacing (Baldwin et al. 2000, Thomas et al. 1995). However, thinning may have more impact than initial spacing (Baldwin et al. 2000). Whether a tree is suppressed or dominant also affects the biomass as a function of diameter at breast height (Naidu, DeLucia, Thomas 1998). As shown in Appendix 6, trees with the same diameter at breast height can have very different heights and correspondingly different biomass. Height differences can arise from differences in both site productivity and tree spacing. The tree population selected for sampling should reflect variation in stand histories and site productivity as much as possible.

Choosing the Categories to Address

The main concern in developing tree biomass equations is choosing which species, size range, and geographical range to address. Quantifiers may want to develop one equation to cover species with similar densities and growth forms.

For woody debris, categories are a key issue. The best approach is to use the five decay classes described in Appendix 10, along with cross sections of tree boles that do not decompose. A two-class system of hard and soft is best for species and systems that decompose relatively evenly across the cross section of the bole.

The process described here assumes that quantifiers are developing biomass equations for a single species. If they are developing an equation for multiple species, they should rely on prior knowledge or pilot testing to ensure that the species have similar growth forms and wood densities. Quantifiers should also sample more trees than they would if developing an equation for a single species. They could estimate the variability in density from the literature and use this variability and equations from a text on regression modeling to estimate the number of trees to sample to obtain the desired level of precision. However, this process is somewhat complex. (See Johnson 2000 and Wharton and Cunia 1987.)

Quantifiers should use sample trees well distributed across the range of diameters to which the biomass equation will apply. Because large-diameter trees occur much less often than small-diameter trees, a simple random sample of all trees would have to be very large to yield enough large-diameter trees. The best approach is to use a version of stratified sampling (see Appendix 1): stratify by diameter and limit the sampling to one or a few trees per stratum. This approach is appropriate because the goal is to determine the relationship of biomass of an individual tree to its height and diameter, not to predict the biomass of a stand or the total mass of trees within a particular area. If productivity varies significantly across the project area, sampling should also be stratified by productivity. To limit the number of trees that must be sampled, quantifiers can divide site productivity (or stand density during development) into only two strata.

To choose strata divided by diameter, first define the minimum and maximum diameters to be encompassed by the equation. The minimum diameter should be 5 centimeters diameter at breast height to allow tighter-fitting of equations. As noted, if resources allow sampling of more trees, quantifiers may want to sample young trees with rapid height growth per unit of diameter growth separately from mature trees with little height growth relative to the same increment of change in diameter. If quantifiers need an equation for small trees, they may want to create one equation for smaller trees and a second equation for larger trees.

Next, determine the total number of trees to sample. The minimum is 30 trees, but at least 50 trees is best (Aldred and Alemdag 1988, West 2003). (See Johnson 2000, Wharton and Cunia 1987, Aldred and Alemdag 1988, and Parresol 1999 for equations for estimating confidence intervals for different forms of regression equations.) Select the number of trees to be sampled in each size class. Divide the total number of trees to be sampled by the number of trees sampled in each size class to yield the number of diameter classes.

To determine the range of each diameter class, subtract the minimum diameter to be sampled from the maximum diameter to be sampled. Divide the range of diameters to be sampled by the number of diameter classes. This yields the span of each diameter class. The smallest size class would extend from the minimum diameter to the minimum diameter plus the size-class span. The next-smallest size class would extend from the upper boundary of the smallest class to that size plus the size-class span. Repeat the process to define all size classes to be sampled. If the equation will apply only to young and moderate-age forests, the largest size class should be a fixed range, like all the smaller size classes. If the equation will apply to an old-growth forest that may have very rare trees of very large size, the largest size class should be open ended, to allow sampling of a very large tree if one is encountered.

Rounding the span to an even number is helpful. Rounding down requires an increase in the number of sampled trees to maintain the desired number of trees in each diameter class. If rounding up, additional trees must be sampled in some size classes to reach the desired total of sampled trees. Sampling more trees in a particular size class gives it more weight when constructing a regression equation. As a result, when more trees are sampled in some size classes, those classes should contain the largest proportion of biomass, to minimize total error in predictions across a project area.

When considering the potential range of tree diameters, quantifiers should also identify the height of a tall tree of the species and size range that will be sampled. They will use this height later to separate sampled trees, to limit correlation between sampled trees.

Identifying Sampling Locations and Trees

Landowners are often unwilling to cut trees to develop biomass equations. Gaining permission to sample requires either paying for destroyed trees or convincing landowners that the research is worthwhile.

The goal of having a relatively small sample that represents the range of trees to which the biomass equation will be applied guides the choice of method for selecting specific trees to sample. Regression assumes that all the points being regressed are independent. Because adjacent trees interact, sampling adjacent trees violates this assumption. Some early studies sampled all trees on a plot. This would be efficient for developing a stand-level equation. However, the goal is to develop equations for individual trees. Thus sampled trees should be at least as far apart as the height of the tallest tree. Ideally, quantifiers would sample only one tree per stand.

To avoid bias, trees must be sampled randomly or through a method that approximates random sampling (see Appendix 1). Totally random sampling would require identifying the diameter class to which each tree belongs and then randomly selecting trees from each class. For equations that will apply to a large area, this would require an impractical amount of work and the availablity of all trees for destructive sampling. In practice, access limits and conservation requirements may make some trees unavailable.

Methods that choose trees using a random start and sample plots along a transect yield results similar to strict random selection (Avery and Burkhart 1994). It is desirable to consider the range of sites to which the equation will apply and classify lands as high or low productivity. Sampling high-productivity sites separately from low-productivity sites will give a range of heights for each diameter sampled. The set of high-productivity areas and low-productivity areas become separate pools for selecting individual trees to sample. If site productivity is relatively homogeneous, quantifiers may decide to sample stands with wide tree spacing separately from stands with narrow tree spacing. This division would require expert judgment, as the sampling necessary to quantify the effects of past spacing would be costly. Quantifiers combine data on all trees for regression analysis.

Consider the likely proportion of hollow trees or stems with substantial decomposition. If most trees in a diameter class are sound, stems in which more than 2 percent of the bole volume is missing or decomposed should be rejected. However, if substantial decomposition is the norm in large trees, quantifiers may decide to sample whatever tree is initially selected, regardless of decomposition. In that case, nonlinear regression will be needed to fit a curve to information on tree weight, and the predictive value of the biomass equation will be lower than if substantial decomposition were not present. If the project area includes a significant number of hollow trees, quantifiers should measure the sizes of hollows using destructive sampling, establish guidelines for characterizing the size of hollow areas, and include the characterization in the protocol for measuring change in tree biomass.

Within accessible lands, randomly choose twice as many starting points as there are trees to be sampled. Randomly choose a direction for each starting point. Order the randomly selected points in a sequence such that it is relatively efficient to travel from one starting point to the next.

During fieldwork, technicians should proceed to the first randomly selected starting point. Consider a plot having a horizontal radius equal to half the height of the tall tree identified when choosing the diameter range to sample. Identify the largest-diameter class in which more trees are needed. Start searching the plot for a tree in this class. Sample the first tree encountered that is the selected size, rejecting the tree only if it has a broken top or if it violates any limits on the degree of decomposition. Do not reject the tree because it is not vigorous or because it has less-than-ideal growth form. The goal is to develop an equation to estimate the biomass of a range of trees, not an ideal tree. If a tree is found and sampled, move to the next plot.

If no tree of the first size class of interest is found on the plot, determine the next-largest size class for which trees need to be sampled. Sample the first such tree encountered that does not have a broken top. If a tree is

encountered and sampled, move to the next plot. If no tree such tree is found, determine the next-largest size class for which trees need to be sampled and search for such a tree. Repeat this process until a tree is sampled or no more size classes remain. If a tree has been sampled, go to the next starting point and start the search process again.

If no tree was sampled at the plot centered on the starting point, then find the randomly selected direction assigned to the current plot. Travel in that direction for a distance equal to the identified height of a tall tree. If this next plot center is outside the boundary of the area to be sampled, stop and go to the next starting point. If the plot center is within the area to be sampled, determine the largest size class for which another sample tree is needed. Search for an appropriate tree and sample the first such tree encountered or move in the selected direction if no appropriate tree is found. Repeat this process until a tree is found and sampled, sampling is complete, or the edge of the plot is reached. If the edge is reached before all needed trees have been sampled, go to the next starting point and continue sampling. It is acceptable if a plot overlaps a plot associated with a previously surveyed starting point.

Traveling to new starting points takes time, and thus exacts a cost, but sampled trees must represent the range of trees and situations to which the biomass equation will be applied. The amount of travel time is also modest relative to the amount of time needed for sampling. This process will involve traveling to at least as many starting points as there are trees to be sampled.

Instead of random sampling, quantifiers may use a systematic sample. This approach to locating plots is best suited to situations where only one or a few large blocks of trees are available for sampling. Divide the area available for sampling using a rectangular grid such that the number of intersections in the grid is roughly twice the number of trees to be sampled. Then randomly choose a starting point for sampling, apply the grid so an intersection is located at the starting point, and move systematically through the points represented by grid line intersections. All needed trees may be sampled before all grid points are visited, and that is acceptable. (See Avery and Burkhart 1994 and other forest measurement texts for more on how to set up a systematic sampling grid.)

The process of choosing specific trees to sample assumes that large trees are rarer than small trees, and it thus gives priority to seeking and sampling large trees. If the selected height of a tall tree is greater than 25 meters or so, several hectares worth of plot area will be available for looking for the largest trees before all the preselected starting points have been searched. If, despite this prioritization, not enough large trees have been sampled when all preselected starting points have been surveyed, field technicians can search for large trees using whatever method they prefer. As long as they sample the first acceptable tree they encounter, bias introduced by nonrandom sampling should be negligible.

Weighing Tree Pieces in the Field

For remote locations, crews can use a block and tackle to lift substantial pieces of tree trunk. With a chain saw and rope, they can cut saplings at the site and fashion them into a tripod for holding the block and tackle. Crews can place the tripod over large pieces, avoiding the need to move the pieces horizontally. If heavy equipment is not available, trees can be cut into small pieces for weighing, and labor can be substituted for machinery. If the trees are valuable, it is probably worthwhile to arrange for equipment that can lift merchantable logs, to allow the sale of the timber.

Sampling Trees in the Field

Equipment for sampling trees in the field includes a chain saw, loppers, tarps, a large-capacity scale that reads to three significant digits, a 10-kilogram scale that reads to three significant digits, and permanent markers. Tarps are useful for weighing branches, sawdust, and any small tree parts that fall off larger parts. Supplies include data sheets and large, sealable plastic bags for subsamples. If sample sites are accessible to wheeled or tracked equipment, a log loader (or any other piece of equipment suitable for lifting intact logs) and load cell for measuring force can dramatically speed fieldwork. If crews use a loader, they will need a chain or equivalent to attach the load cell to the loader's lifting apparatus. Chokers are also useful for lifting logs. Make sure the hooks on chokers can be attached to the eye on the load cell.

If a loader is not available, crews will have to cut trees into pieces small enough to lift by hand. A peavey, burly field technicians, and a lever lifting system make the process more tractable. If the weighing device can easily tare to zero weight, tare the device while lifting the empty tarp or cable, and then lift and weigh the tree piece. If the weighing device does not easily tare to zero, for each piece weighed, record the gross weight of the tree piece, including whatever tarp or cable is used to lift it, and record the weight of the tarp or cable not including any tree part. Subtracting to find net weights in the field invites errors, so that should occur in the office.

Ideally, all sampling should be done when trees are not wet from precipitation, to limit variation in dry-to-wet weight ratios of subsamples. Dry material is more important for foliage measurements than for bole measurements. In some climates and with some species, water content will vary measurably by time of day and season. This variation will increase the variance of the observed dry-to-wet weight ratios.

The process for sampling and weighing trees depends on whether they are to be harvested for lumber. Large trees are valuable, and destructive sampling may be available only if it leaves logs suitable for sale. This would require a load cell or other device that measures up to several tons. A 10-meter-long log with a midpoint diameter of 0.8 meters (31.5 inches) can easily weigh 5 metric tons, and a 10-meter-long log with a midpoint diameter of 1.2 meters (48 inches) can easily weigh 10 metric tons. If power equipment is not available for weighing logs, pieces will have to be cut small enough to be lifted for weighing.

After selecting the tree to be sampled, note on the list of trees to be sampled that a tree of the appropriate diameter class has been measured. Before felling the tree, record the plot coordinates (or other plot location code). Then measure and record the tree's diameter and height. Fell the tree. If the tree will be cut into many pieces, place a tarp to collect the bulk of the sawdust, and weigh the sawdust. Use a measuring tape or laser measuring device to measure the tree height along the felled stem (including allowance for the stump), to check the height measured while the tree was still standing. Record the height measurement from the felled stem.

Crews will take four subsamples from each tree, for determining the ratio of dry weight to wet weight. One subsample will be branch and foliage material, and three will be stem bark and wood. This ratio is roughly proportional to the masses of branches plus foliage to stem mass in most trees.

To determine where to take subsamples, first measure the length of the live crown. Generate a random number between zero and one to three decimal places and multiply it by the length of the live crown. Starting at the base of the live crown, move toward the tree tip the distance calculated by multiplying the live crown length by the random number. Collect the subsample at this point. For example, if the live crown is 10 meters long and the random number is 0.675, crews would measure 6.75 meters up from the base of the live crown. Select the branch that is closest to the subsampling point. If the closest branch is in a whorl of branches, select the branch that is closest to pointing straight up in the air. Weigh this branch with its attached foliage and dead twigs, and record this weight on the section of the data sheet for total branch weights.

During data analysis, quantifiers will sum all branch weights and stem weights. They will multiply each of these sums by the dry-to-wet weight ratio for the respective components and add the dry weights to yield total dry weight of the aboveground biomass of the tree.

If the branch is too large to conveniently bag, transport, and dry, cut twigs (with foliage) off the main stem of the branch and place all the twigs and foliage on a tarp. Then cut the main stem of the branch into short sections of about equal length, select and retain every third section, and discard the rest. Divide the twig and foliage material into three equal-sized piles. Select one of the three and discard the rest. Combine the retained twig and foliage material with the retained branch stem material. If this still does not reduce the mass of branch material to a manageable amount, divide each pile in half again and discard half. Although it is best

if the average wet subsample weight is no more than 2 kilograms, some subsamples can be up to 10 kilograms, and many subsamples can weigh about 1 kilogram. After creating a subsample of acceptable size, do the following: tare a scale to the bag in which it will be stored; place the subsample in the bag; seal and weigh the bag; and write on the bag the tree number, the label of branch subsample, and its field wet weight. Retain the subsample for processing.

Next, cut all branches from the bole of the tree, place them on tarps, and weigh them. Record the weight (or weights if there is too much material to weigh at one time). If the weights include tarps or other containment or lifting material, record the weights of this material next to the gross weights. If the scale is tared to zero while the tarp (or other containment material) is lifted, and the recorded weights include the tree material, then record zero in the column for the lifting apparatus weight.

Cut off any stump at ground level. Weigh the stump, record the weight as stump weight, and record the weight of any lifting apparatus that quantifiers will have to subtract from the measured weight to find the weight of the tree stump.

If the tree trunk does not need to be cut into specific lengths for timber, cut the trunk into thirds of equal length. Weigh the trunk, or each trunk section if more convenient, and record the weight. Also, record the weight of any lifting apparatus, which will be subtracted from the measured weight to find the weight of the tree material.

From the large-diameter end of each tree trunk section, cut a disk of wood 2 to 4 centimters thick. If a disk weighs less than 1 kilogram, bag the entire disk. If the disk weighs 1 to 2 kilograms, cut the disk into two half-circles and retain one half. If the disk weighs more than 2 kilograms, cut a wedge of wood shaped like a slice of pie so the point of the wedge is located at the pith of the stem and the wide part of the wedge is the bark. The goal is to obtain a subsample that weighs no more than 2 kilograms. Very thin wedges are hard to cut accurately, and for large-diameter stems, the subsamples may weigh more than 2 kilograms. Tare the scale to the bag weight and individually weigh each subsample. Write the tree number, sample location (stem bottom, middle, or top), and field wet weight on each sample and bag. If samples are broken to fit into a bag, write the sample information on each piece. Include all bark. Record the field wet weight of each sample.

If the tree trunk is to be bucked into specific lengths for timber, leave an extra 3 centimeters in the waste allowance of each log, for cutting a subsample from between each log. Buck the tree, weigh each piece, and record the weights prior to subsampling.

If the tree yields only one log, cut two wedge samples from opposite sides of the top of the stump and one subsample from the base of the tree tip section. If the tree trunk yields two logs, cut one subsample from the stump, one from the top of the first log or the bottom of the second log, and one from the bottom of the tree tip segment or the top of the second log. If the tree yields three or more logs, cut one subsample from the stump, one from the top of the first log or bottom of the second log, and one from the top of the top log or bottom of the tree tip segment. Follow the same procedure for sizing, cutting, weighing, and labeling subsamples described above.

Processing Subsamples

As soon as possible, samples should be dried at 65°C until their weight is constant. This usually takes several days, longer for branch segments that are longer than a few centimeters. Immediately after drying, weigh and record the dry weight of each subsample. If the subsample is in pieces, weigh all pieces together.

Analyzing the Data

To analyze all the field data, first calculate and inspect the dry-to-wet weight ratios of each subsample, examining any extraordinary values to see if they are erroneous. If any are incorrect, try to correct them using the redundancy in data recording or other means. If correction is impossible, discard the ratio for that subsample. Next, calculate an average value and standard deviation for the wet-to-dry ratios of all branch/foliage subsamples. Because only a single branch/foliage subsample is measured from each tree, quantifiers cannot do statistics to determine the sampling error. Instead, they should multiply the field wet weight of the branches and foliage of each tree by the average dry-to-wet weight ratio for the branches and foliage of all trees sampled. This yields a dry weight estimate for the foliage and branches of each tree sampled. The standard error of the estimate can be calculated and reported for the average dry-to-wet weight ratio of all trees sampled.

Next, for each tree sampled, find the average of the dry-to-wet weight ratios of the three subsamples. Multiply the ratio for each tree by the measured green weight of the tree bole to calculate the dry weight of the tree bole. For each tree, sum the dry weights of the stem, foliage, and branches to calculate the dry weight of the total aboveground biomass.

Analysts have used a variety of forms of regression equations to predict tree biomass as a function of height and diameter. Aldred and Alemdag (1988) and Parresol (1999) assess the strengths and weaknesses of several model forms.

Because tree density is relatively constant across tree sizes, quantifiers can use equation forms developed for calculating tree volume to calculate tree biomass (Aldred and Alemdag 1988). Tree biomass equations are more directly applicable to measuring carbon stocks than tree volume equations. Quantifiers must convert the results of volume equations to mass by multiplying by wood and bark density, and this gives more opportunity for error. Volume equations are also harder to create than simple biomass equations. (For a detailed description of how to develop tree stem volume equations and forms of equations, see West [2003].)

Crown biomass is the most variable component of aboveground biomass. Measuring the live crown length can slightly improve estimates of total aboveground tree biomass, but the improvement is so small that it is not worth the effort (Aldred and Alemdag 1988).

Typical biomass equations forms are

$$M = a \times D^2 \times H,$$
$$M = a + (b \times D^2 \times H),$$
$$M = a \times D^b \times H^c$$

Equation A14.1

where M is the aboveground biomass of the tree; a, b, and c are regression coefficients, D is the diameter at breast height, and H is the total tree height.

Each equation is developed for specific units of mass, diameter, and height. Equations with y-intercept terms often give inaccurate predictions at the small end of the diameter range. However, because the total biomass of small trees is small, these errors will not significantly affect the outcome (Aldred and Alemdag 1988).

Biomass equations often use the logarithms of diameter, height, and biomass for two reasons. First, tree biomass does to not grow linearly with diameter and height, and the relationship of the logarithms of these variables is substantially linear. Second, regression requires that the errors be constant across the range of the equation, but untransformed values produce larger errors at larger diameters. Log transformation makes errors constant across the range of values. Base 10 logarithms have been used in transformations, but the natural logarithm (base e) transformation seems to be favored for conducting ordinary least squares regression.

When log-transformed equations are used, back-transforming with the antilog to yield biomass estimates in terms of the original units underestimates biomass. Quantifiers can correct the estimate by a factor that includes the variance about the logarithmic regression line. (Johnson [2000] and Wenger [1984] provide such factors; see Parresol [1999] for a more complete discussion.)

Quantifiers can avoid this correction by using a weighted equation instead of a nonlinear log-transformed equation. When variance of errors changes as a function of a variable or variables, weighting by a function of that variable can make the variance constant over the observed range. Weighted equations can give tight confidence intervals around biomass estimates, without uncertainty about whether the antilog bias correction is accurate (Parresol 2001).

For small trees, the regression model recommended here is

$$M = b \times D^2 \times H \qquad\qquad \text{Equation A14.2}$$

where M is the aboveground biomass of the tree, b is a regression coefficient, D is the diameter at breast height, and H is the total tree height.

This model has geometric correspondence to the volume of the bole of trees. Because it passes through the origin, it can be fitted to give good results for small-diameter trees with a positive diameter at breast height (Aldred and Alemdag 1988). Because the bulk of tree biomass is in the bole, it is desirable to use an equation that reflects tree bole geometry.

For larger trees, quantifiers can obtain a somewhat better fit using the following equation:

$$M = b_0 + (b_1 \times D^2 \times H) \qquad\qquad \text{Equation A14.3}$$

where M is the aboveground biomass of the tree, b_0 and b_1 are regression coefficients, D is the diameter at breast height, and H is the total tree height (Parresol 1999). This model has heteroscedastic variance, and errors should be modeled. Errors cannot simply be reported using the fit index r^2.

A generic equation for stem wood volume that will return a volume estimate that is within about 15 percent of the correct value for many species is

$$V_U = 0.281 D^{1.91} \times H^{1.02} \qquad\qquad \text{Equation A14.4}$$

where V_U is the volume in cubic meters of stem wood (excluding bark) from ground to upper tip, D is the diameter of the tree at breast height (over bark) in meters, and H is the total tree height in meters (West 2003). This equation was created by averaging volume predictions of equations developed for individual hardwood and softwood species from around the world.

Provisional equations should be checked to ensure that predicted biomass does not decrease as diameter increases (except possibly for trees that become hollow at large diameters). Quantifiers should also check that errors are normally distributed and that the equal-variance rule is not grossly violated. Inspection of a scatter plot of errors can reveal nonlinearity. Errors can be weighted by the factor of $1/(D^2 \times H)$, and after weighting, the errors should show equal variance across the range of tree sizes addressed by the equation (Aldred and Alemdag 1988). If a provisional equation fails any of these tests, additional fitting should be performed.

Validating the Equations

Ideally, factors should be validated by applying them to data other than those from which they were developed. However, such data usually do not exist because that is why new equations were developed.

Still, quantifiers can collect additional data for use in validation. For tree biomass equations, tree field weight can be measured, and the ratio of dry-to-field weight can be developed in the research used to calculate their dry weight. This is still significant work, but less than the complete set of measurements used to construct equations.

If validating, quantifiers use points near the end of the range to which the equation applies, especially the end where most sequestration occurs. For tree biomass equations, this means using some large trees for validation. The equations should be accurate where the largest amounts of sequestration will occur.

An alternative to collecting field data for validation is comparing predictions from the new equations to existing equations. These include the generic equations above, as well as the equations and sources of equations included in the section on writing monitoring plans (see Chapter 2). Results from the new equations obviously will not match results from existing equations, or it would have been fine to use the existing ones. However, the differences should vary in a predictable way.

Appendix 15

Using Stand-Level Equations

Stand-level equations predict carbon stock per unit of area instead of per single specimen. Biomass equations for individual trees are usually more accurate than stand-level equations. However, if detailed data on individual specimens and their carbon content may be too expensive to obtain, stand-level equations may be the best option. Stand-level equations are most reliable when applied to stands dominated by trees of relatively uniform size.

Most stand-level equations predict carbon mass as a function of tree species and the average diameter at breast height of trees in the stand. Biomass expansion factors, on the other hand, predict total aboveground biomass as a function of merchantable timber volume.

When limited data on forests are available, that information may include the dominant species and average diameter of trees in the stand. If not, these aspects are relatively inexpensive to measure. However, stand-level equations are confounded by variation in site productivity, and they do not account for different stand histories, which affect the amount of carbon storage in woody debris and litter.

Tree density affects carbon stock in two ways that a simple stand-level equation does not address. First, if a stand has fewer trees than average, even if each tree contains the same amount of carbon as all others of the same diameter, that less-dense stand will have less carbon per unit area than a denser stand. Second, on any given site and for any given diameter, trees that grow more densely tend to be taller than trees that grow less

densely. Thinning can have an even greater effect on biomass as a function of diameter than tree spacing at stand initiation (Baldwin et al. 2000, Naidu at al. 1998, Thomas et al. 1995). Because of these dynamics, stand-level equations based on average stand diameter can yield biomass estimates that are 30 to 40 percent larger or smaller than estimates made using individual tree diameters (Jenkins et al. 2003). Whether such large potential errors are acceptable will depend on the regulatory or market system.

Another approach to estimating stand-level carbon stocks is to use a biomass expansion factor to estimate carbon mass as a function of merchantable timber volume and forest type. Such expansion factors are larger for lower timber volumes. This reflects typical stand conditions, in which lower volumes mean smaller trees, which have a higher proportion of biomass in nonmerchantable parts. As with stand-level equations, biomass expansion factors are easy to apply to projected or cruised timber volumes. Like other stand-level equations, biomass expansion factors apply an average value that integrates the impact of management and disturbance histories. As a result, for some stands, the biomass estimated by calculating the mass of individual trees and woody debris can differ markedly from the biomass calculated using a biomass expansion factor. The largest differences tend to occur in stands with lower volumes of growing stock (Smith, Heath, and Jenkins 2003).

Appendix 16

Calculating Changes in Carbon Sequestration When Soil Density Changes

Quantifiers should calculate the soil carbon present at each sampling site at the start of a project. If carbon in roots is also of interest, quantifiers should use the same protocol to calculate that stock. For each site, inputs to the calculation of carbon stock per unit of area include

- The bulk density of the fine soil sampled.
- The volume of fine soil sampled.
- The proportion of fine soil that is carbon, by weight.
- The cross-sectional area sampled.

The dry mass of fine material and bulk density are measured during sample processing. The cross-sectional area sampled at each site is calculated by multiplying the cross-sectional area of a single core by the number of cores collected at each site:

$$SampleArea = NumbCores \times (\pi \times (CoreRadius^2))$$

Equation A16.1

where *SampleArea* is the total area sampled at that site in square centimeters, *NumbCores* is the number of cores sampled at each site, and *CoreRadius* is the radius of each core in centimeters.

For example, if 16 cores are sampled at each site, with each core having a radius of 3 centimeters, the area sampled is

$$SampleArea = 16 \times (\pi \times (3^2)) = 16 \times 28.2744 =$$

$$452.390 \text{ cm}^2$$

The proportion of fine soil that is carbon is measured by analysis of dry combustion. The fine soil volume, bulk density, proportion of the mass that is carbon, and sample mass are converted to tons of carbon per hectare[1]:

$$MgC/ha = \frac{FineSoilVolume \times Density \times ProportionC \times 100}{SampleArea}$$

Equation A16.3

where *MgC/ha* is megagrams (millions of grams, or metric tons) of carbon per hectare; *FineSoilVolume* is the total volume of fine soil sampled at that site in cubic centimeters; *Density* is the bulk density of the fine soil in grams per cubic centimeter; *ProportionC* is the proportion of the fine soil that is carbon, by weight (not percentage) as determined by dry combustion and number of cores sampled at each site; and *SampleArea* is the total area sampled at that site in square centimeters. The factor 100 converts grams per square centimeter to megagrams per hectare.

Continuing the example, suppose that sampling occurs to a depth of 20 centimeters and the soil contains no rocks or roots. Fine soil volume is 9,047.8 grams, the bulk density is 1.3 grams per cubic centimeter, the proportion carbon is 0.01 (which is 1 percent carbon, or about 1.72 percent organic matter), and the sample area is 452.39 square centimeters. The carbon stock would be

$$MgC/ha = \frac{9,047.8 \text{gm} \times 1.3 \text{gm}/\text{cm}^3 \times 0.01 \times 100}{452.39 \text{cm}^2} = 26$$

This is the amount of carbon per hectare to the depth sampled. Sampling to a different depth would yield a different reading of soil carbon. If the analysis is in units other than grams and centimeters, quantifiers will need different conversion factors.

Calculating Soil Density

Quantifiers calculate bulk soil density for each site in grams per cubic centimeter. Bulk density is used to check for changes in soil carbon density over time, and it can be used to calculate carbon stocks on a per-unit-of-area basis directly. To reduce the number of calculations and the chance of error, the approach here uses the inputs to the bulk density calculation (the dry weight and volume of soil samples) to calculate carbon stocks on a per-unit-of-area basis.

The bulk soil density for each sampling site is the rock-free density of fine soil that passes through a sieve with 2-millimeter openings without grinding.[2] The dry mass of soil from each site is divided by the volume sampled at that site. The volume is the sum of the volumes of all cores combined at that site. The equation for calculating the volume is

$$Volume = NumbCores \times \left(CoreLength \times \left(\pi \times \left(CoreRadius^2 \right) \right) \right)$$
Equation A16.5

where *Volume* is the total volume of the cores sampled at that site in cubic centimeters, *NumbCores* is the number of cores sampled at each site, *CoreLength* is the length of each core in centimeters, and *CoreRadius* is the radius of each core in centimeters.

This equation assumes that all cores are taken to the same depth. If that is not the case, quantifiers must calculate the average core length. Finding the bulk density of fine soil requires accounting for the volume of rock fragments and woody material in the samples. Quantifiers must subtract the volume of rock fragments (fragments not ground and added to the fine soil) and wood from the total sample volume to find the volume of fine soil.

The dry masses of rocks and woody material are found using laboratory analysis. The masses can be converted to volume using default densities or densi-

ties measured in the samples. The default density for rock fragments is 2.65 grams per cubic centimeter, and the default density for woody material is 0.5 grams per cubic centimeter. Volume is calculated as

$$Volume = \frac{Mass}{Density}$$
Equation A16.6

Volume is calculated separately for rocks and woody material. The volume of fine soil is

$$VolumeFineSoil = TotalSampleVolume - (RockVolume + WoodVolume)$$
Equation A16.7

To find the bulk density, divide the dry weight of fine soil (in grams) by the volume:

$$BulkDensity = \frac{DryWeight}{VolumeFineSoil}$$
Equation A16.8

Correcting for Change in Bulk Soil Density

To calculate changes in fine soil density, quantifiers should use samples collected for measuring soil carbon, not the deeper increments available for correcting for changes in soil density:

$$\Delta FineSoilDensity = FineSoilDensity_2 - FineSoilDensity_1$$
Equation A16.9

where Δ is the change, subscript 2 refers to the later remeasurement, and subscript 1 refers to the initial measurement. The change will be positive if the density is rising, and negative if density is falling.

To calculate the change in the mass of fine soil from one sampling to the next, quantifiers should use data from the same cores:

$$\Delta FineSoilMass = \Delta FineSoilDensity \times Volume_1$$
Equation A16.10

If the change in fine soil mass is more than 1 to 2 percent of the total mass, quantifiers should account for it. This requires using the proportion of carbon, by weight, from the deeper samples:

$$DensityCorrection = \frac{\Delta FineSoilMass \times ProportionC \times 100}{SampleArea}$$
Equation A16.11

where *ProportionC* is the proportion of carbon measured in the extra depth increment of soil collected for correcting for change in bulk density.

The density correction is in metric tons of carbon per hectare, the change in fine soil mass is in grams, and the area sampled is in square centimeters. The proportion of carbon is unitless. The factor 100 converts from grams per square centimeter to tons per hectare.

The change in carbon stock at each sampling site is calculated as

$$\Delta Carbon_{site} = MgC/ha_2 - MgC/ha_1 - DensityCorrection$$

Equation A16.12

All the elements of this equation are in units of metric tons of carbon per hectare.

Appendix 17

Determining Mass-Specific Ratios

The mass-specific ratio is the average ratio of the dry weight of manure samples to their wet weight or volume. The process of determining that ratio involves collecting samples, measuring their weight or volume, drying them, and measuring the dry weight of each sample. Quantifiers calculate the ratio for each sample and then the average ratio and its uncertainty (see Appendix 3 on statistics). Samples can be dried in an oven at a temperature between 60°C and 105°C, or water can be removed from liquid samples by boiling or microwaving. Heating results in the loss of some volatile solids, but the loss should be modest. Samples should be processed within several hours of collection to avoid loss of organic material through decomposition. If samples cannot be processed immediately, they can be frozen or sterilized for storage. After drying, immediately measure and record the dry weight.[1]

Determining Ratios by Weight

Wet-weight samples should be collected at several different times, spanning the range of manure types that will occur in the project. Enough samples should be collected to ensure that the average ratio calculated from the samples is accurate. The average specific ratio should have no more than a 10 percent uncertainty at a 95 percent confidence level (see Appendix 3). Collect about 15 samples and determine if they provide adequate accuracy. If they do not, collect and analyze more samples, pool the new data with the earlier data, and recalculate the confidence interval.

Analyzing samples is simple for systems that handle manure in solid form. Weigh the empty sample containers, add wet manure to each container, record the wet weight of each sample with the container, and then dry the samples and weigh again. For solid, relatively dry manure, samples of about 1 kilogram are easy to handle and large enough to average out some of the variability in the manure. They also are large enough to allow the use of inexpensive balances to obtain accurate weights. As a general rule, the measurements used to calculate project-specific ratios should be made to at least three significant digits.

To find the average dry-to-wet-weight ratio, first find the ratio for each sample:

$$R_n = \frac{dry_n}{wet_n} \qquad \text{Equation 17.1}$$

where R_n is the ratio for sample number n, dry_n is the dry weight of sample n, and wet_n is the wet weight of the sample n. The two weight terms, wet_n and dry_n, must be in the same units, such as grams.

Then add the ratios and divide by the number of samples:

$$R_{average} = \frac{\sum_{1}^{N} R_n}{N} \qquad \text{Equation 17.2}$$

where $R_{average}$ is the average ratio of dry weight to wet weight, and N is the number of samples.

Determining Ratios by Volume

For liquid manure, the method presented here assumes that the manure flow is measured by volume rather than weight. The goal is to get a density of the dry weight per unit of volume. A sample size as large as 1 liter may be necessary to capture a representative mix of solids and liquid. Be sure to take samples that are representative of the manure stream. For example, if solids are washed into a digestion tank, capturing samples as the slurry enters the tank will be more representative than taking samples from the surface of the tank after the solids have begun settling out of the liquid. Try to collect samples that match the distribution of manure conditions. If about 10 percent of the manure is much wetter than the other 90 percent, collect 10 percent of the samples in the wet manure. Carefully measure and record the volume of each sample, and then dry the samples and measure the mass or weight of the dried samples.

The volume of liquid samples is easily measured using a large graduated cylinder. The volume of solid manure is somewhat more difficult to measure because of the irregular shapes of pieces of solid manure. If the ratio of dry weight to volume of manure is needed, it may be easiest to measure the volume and wet weight of a large volume of solid manure, such as a container of several dozen to a few hundred liters. This relatively large amount of manure can be shoveled into a container whose volume can be measured fairly accurately. Then the manure can be subsampled, and the volume of the subsample can be inferred from its weight. The subsample can then dried and weighed to find the ratio of dry weight to volume.

For example, suppose that a volume of 100 liters of manure is sampled, and that this mass weighs 50 kilograms. If a representative subsample of 0.5 kilograms is taken from the sample, this subsample can be assumed to have the same density as the sample as a whole and thus can be assumed to be 1 liter in volume. The subsample is then dried, and the ratio of dry mass to volume is found for the subsample.

The average ratio is found by calculating the ratio for each sample, adding all the ratios, and dividing by the number of samples. So, first calculate the ratio for each sample:

$$R_n = \frac{DryWt_n}{WetVol_n} \qquad \text{Equation 17.3}$$

where R_n is the ratio for sample number n, $DryWt_n$ is the dry weight of sample n in grams, and $WetVol_n$ is the wet volume of the sample n in cubic centimeters. Then add the ratios and divide by the number of samples:

$$R_{average} = \frac{\sum_{1}^{N} R_n}{N} \qquad \text{Equation 17.4}$$

where $R_{average}$ is the average ratio of dry weight to wet volume, in grams per cubic centimeter, and N is the number of samples. This method gives each sample equal influence on the calculation of the average, even if the samples have different weights.

Finally, the uncertainty in the average ratio is calculated following the methods described in Appendix 3. We recommend that the estimate of the ratio have a statistical precision such that the 95 percent statistical confidence interval is no more than ±10 percent of the mean value of the ratio.

If manure-handling practices change, it may be necessary to develop a new dry-to-wet-weight ratio. For example, changing the equipment used to flush animal pens may change the ratio of water diluting the manure.

Appendix 18

Calculating Methane and Nitrous Oxide Emissions

Methane

Quantifiers can estimate methane emissions as follows:

$$Methane_{CO_2e} = \left[\sum_{type,treat} \left(MM_{type,treat} \times P_{type} \times TF_{treat} \right) \right] \times GWP$$

Equation A18.1

where $Methane_{CO_2e}$ is the CO_2 warming equivalent of the direct methane emissions of the project for the period, $MM_{type,treat}$ is the dry-weight manure mass of each type of manure input into each manure treatment system, P_{type} is potential mass of CH_4 that could be generated from manure of each type, TF_{treat} is the proportion of potential CH_4 emissions that results from each treatment, and GWP is the global warming potential of methane.[1]

If project emissions are not measured, quantifiers can estimate project emissions for selected types of manure management systems using values of P_{type} in Table A.16 and values for TF_{treat} for selected manure treatment systems from Table A.17.

For example, consider a project where 40 percent of the manure is spread daily and the remaining 60 percent is placed in an open anaerobic lagoon. Assume that the project produces 1,000 metric tons of manure over an accounting period. Using the appropriate numbers in Equation A18.1 yields a methane emission over the accounting period of 2,033 metric tons of carbon dioxide equivalent (see Table A.18).

Nitrous Oxide

Quantifiers can use the following equations to estimate the CO_2 equivalent of nitrous oxide emissions from the dry weight of manure, the expected rate of N_2O emissions by manure management system, and global warming potential:

$$MN_{type} = MM_{type} \times N_{ratio}$$

Equation A18.2

where MN_{type} is the total mass in metric tons of nitrogen in each type of manure input into each manure treatment system, MM_{type} is the total dry weight mass in metric tons of manure of a type, and N_{ratio} is the proportion of nitrogen in manure of type MM.

The mass of nitrous oxide emitted in carbon dioxide equivalent can be calculated as

$$NitrousOxide_{CO_2e} = \left[\sum_{type,treat} \left(MN_{type} \times TF_{treat} \right) \right] \times GWP$$

Equation A18.3

where $NitrousOxide_{CO_2e}$ is metric tons of CO_2 warming equivalent of the direct N_2O emissions of the project for the period, MN_{type} is the total number of metric tons of nitrogen in each type of manure input into each manure treatment system, TF_{treat} is the expected mass of N_2O emissions as a proportion of nitrogen input for each treatment, and GWP is the global warming potential of N_2O.[2] TF_{treat} can be measured or taken from published values such as those in Table A.17.

Table A.16 Maximum P_{type}, Potential Methane Production from Manure, Assuming Average Feed

Type	Country or development status	P_{type} (Mg CH$_4$ per Mg manure, dry weight)	
		IPCC	EPA
Dairy cattle	Developed	0.1608	0.15888
Dairy cattle	Developing	0.0871	
Nondairy cattle	Developed	0.1139	
Nondairy cattle	Developing	0.067	
Swine	Developed	0.3015	0.31776
Swine	Developing	0.1943	
Buffalo	All	0.067	
Sheep	Developed	0.1273	
Sheep	Developing	0.0871	
Goats	Developed	0.1139	
Goats	Developing	0.0871	
Camels	Developed	0.1742	
Camels	Developing	0.1407	
Horses	Developed	0.2211	
Horses	Developing	0.1742	
Mule/asses	Developed	0.2211	
Mule/asses	Developing	0.1742	
Poultry	Developed	0.2144	
Poultry	Developing	0.1608	
Cattle, feedlot	USA		0.21846
Turkeys, broilers	USA		0.23832
Other poultry	USA		0.25818

Note: Mg = megagrams, or metric tons.
Source: Calculated from IPCC (2000), EPA (2001), and EPA (2004).

Table A.17 TF_{treat}, Proportion of Potential CH_4 Emissions for Various Types of Manure Treatments

System	TF_{treat}			Notes
	Cool conditions	Temperate conditions	Warm conditions	
Pasture	0.01	0.015	0.02	
Daily spread	0.001	0.005	0.01	
Deep litter	0	0	0.3	Cattle & swine; storage < 1 month
Deep litter	0.39	0.45	0.72	Storage > 1 month
Poultry	0.015	0.015	0.015	
Solid storage	0.01	0.015	0.02	
Dry lot	0.01	0.015	0.05	
Pit	0	0	0.3	Storage < 1 month
Pit	0.39	0.45	0.72	Storage > 1 month
Liquid slurry	0.39	0.45	0.72	
Anaerobic lagoon	0–1	0–1	0–1	Depends on rate of CH_4 capture & destruction
Anaerobic digester	0–1	0–1	0–1	Depends on rate of CH_4 capture & destruction
Composting—extensive	0.005	0.01	0.015	
Composting—intensive	0.005	0.005	0.005	
Aerobic	0.001	0.001	0.001	

Notes: See Chapter 8 for information on management systems. A cool temperature is an average annual temperature less than 15°C; warm is an average annual temperature greater than 25°C; and temperate is an average annual temperature between warm and cool.
Source: IPCC (2000).

Table A.18 Calculating Emissions When the Manure Stream Is Divided between Two Treatment Practices, for a Total Waste Production of 1,000 Tons during the Accounting Period

Treatment	Percentage of waste in treatment	Manure dry mass, Mg	P_{type}, Mg CH_4/ Mg dry manure	TF_{treat}	GWP, CH_4	Emission, Mg CO_2e
Daily spread	40%	400	0.1608	0.005	21	7
Lagoon	60%	600	0.1608	1.0	21	2026
Total	100%	1000	—	—	—	2033

Note: Mg = megagrams, or metric tons.

Appendix 19

The Dynamics of Methane and Nitrous Oxide Emissions from Soil

In soil, chemical transformations of nitrogen (called nitrification and denitrification) can produce methane. The production of methane (called methanogenesis), nitrification, and denitrification occur through electron exchange between two substrates, one an electron donor and the other an electron acceptor. The so-called soil redox potential, a measure of the soil's ability to oxidize a compound, is the most important factor in determining which compound will serve as an electron donor and which will be an electron acceptor. The soil redox potential is denoted with the symbol Eh. The strength of the redox potential is measured in millivolts, denoted mV. Soils with lots of oxygen have a high Eh and will tend to oxidize, or take electrons from, compounds in the soil. Soils that are poor in oxygen (such as those saturated with water) have a low Eh and will tend to reduce, or donate electrons to, compounds in the soil.

Most soil microbes obtain energy by breaking down organic carbon bonds. These processes require the carbon to donate electrons to an acceptor. If a soil is aerated with high Eh (about 600 mV), oxygen (O_2) is always the dominant electron acceptor. By accomplishing the electron exchange, the microbes gain energy and, at the same time, convert carbon (C) and O_2 into CO_2. If soil pores are saturated with water from rainfall or irrigation, water can block the diffusion of O_2 into the soil. With biological consumption of O_2 in the soil, the O_2 concentration will rapidly fall and drive soil Eh downward toward 200 mV or less. Under these O_2-defi-

cient conditions, the soil microbes will have to look for a new electron acceptor to survive. In most soils, NO_3^- is the second candidate electron acceptor.

A special group of microbes, denitrifiers, have the capacity to use NO_3^- as an electron acceptor, provided the soil's Eh is sufficiently low. When the denitrifiers obtain energy, they donate electrons from C to NO_3^-. This process converts NO_3^- to NO_2^-. During the sequential processes of denitrification, the oxidation states of N will reduce from +5 in NO_3^- to +3 in NO_2^-, to +2 in NO, to +1 in N_2O, and finally to 0 in N_2 as the N receives electrons. Thus the ultimate product of denitrification is N_2, which is not a greenhouse gas. However, N_2O, which is a potent GHG, is an intermediate product. Some N_2O often escapes into the atmosphere during the denitrification process, before it can be converted to N_2. This is the reason that adding nitrogen fertilizer to soils leads to N_2O emissions. And this is also the reason why land-management practices that seek to keep soil Eh high can help minimize these emissions (see Table 9.4).

If a soil is flooded for a relatively long time, such as several days or weeks, most of its oxidants (O_2, nitrate, manganese or Mn^{4+}, iron or Fe^{3+}, and sulfate) will be depleted, and the soil Eh will drop to as low as –150 mV. Under these deeply anaerobic conditions, methanogens use H_2 as electron acceptors to produce CH_4. When a wetland soil is drained, the sequence above will reverse, and the dominant process will shift from methanogenesis to denitrification, then further to ni-

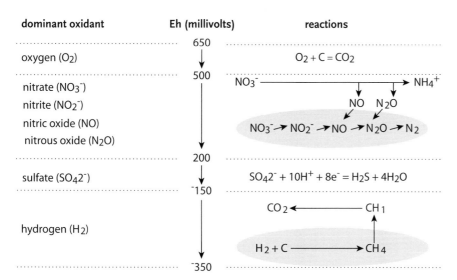

Figure A.13 Trace gas emissions from soil as a function of soil Eh (Li et al. 2004).

trification and CO_2 production. Figure A.13 shows how CO_2, N_2O and CH_4 are produced, driven by soil Eh and substrate concentrations.

Two basic laws of physics and biology can be used to quantify soil redox potential and the resulting microbial activities. The Nernst equation (Equation A19.1) is a classical thermodynamic formula that quantifies soil Eh based on concentrations of the oxidants and reductants in the soil liquid phase (Stumm and Morgan 1981):

$$Eh = E_0 + \frac{RT}{nF} \times \ln \frac{[oxidant]}{[reductant]} \qquad \text{Equation A19.1}$$

where Eh is the redox potential of the soil reduction-oxidation system (V), E_0 is the standard electromotive force (V), R is the gas constant (8.314 J/mol/k), T is the absolute temperature ($273 + t$, where t is the temperature in $^\circ$C) n is the transferred electron number, F is the Faraday constant (96,485 C/mol), $[oxidant]$ is the concentration (mol/l) of the dominant oxidant in the system, and $[reductant]$ is the concentration (mol/l) of the dominant reductant in the system.

Soil microbial activity can be defined by the Michaelis-Menten equation (Equation A19.2). This equation is a widely applied formula describing the kinetics of microbial growth with dual nutrients (Paul and Clark 1989):

$$F_{[oxidant]} = a \times \frac{DOC}{b + DOC} \times \frac{[oxidant]}{c + [oxidant]} \qquad \text{Equation A19.2}$$

where $F_{[oxidant]}$ is the fraction of the oxidant reduced during a time step; DOC is the available C concentration; $[oxidant]$ is the concentration (mol/l) of the dominant oxidant or electron acceptor in the system; and a, b, and c are coefficients.

The Nernst equation quantifies soil Eh based on concentrations of dominant oxidants and reductants existing simultaneously in a soil liquid system, and the Michaelis-Menten equation tracks the consuming rates of substrates driven by the soil microbial activity. Because the two equations share a common term, oxidant concentration, they can be merged to simulate the soil biogeochemical processes driven by the microbiologically mediated redox reactions. Given the concentrations of dissolved organic carbon, oxidants, and reductants, the linked Nernst and Michaelis-Menten equations can be solved to determine the soil Eh status, microbial growth rates, and consumptions of the oxidants and available carbon through which nitrous oxide or methane is produced (Li et al. 2004).

Appendix 20

Market Leakage and Activity Shifting

Some analysts distinguish between two different types of displacement: market leakage and activity shifting. (See, for example, the World Resources Institute GHG project-accounting protocol 2003.)

Market leakage occurs when a project reduces the production of some good supplied to a market, without a corresponding reduction in the demand for that good. Demand is displaced from the project to other suppliers. Chapter 10 describes methods for quantifying leakage resulting from this type of displacement.

Activity shifting is the displacement of activities to other locations when project lands are no longer available for use. Activity shifting is relevant when there is a significant subsistence use of lands. For example, local people might extract fuel wood from a forest for domestic use. If a project prevents them from continuing to use the forest as a source of fuel wood, they will obtain it elsewhere.

This book considers such activity shifting a non-cash displacement of production and thus a market displacement. That is because, even if local people were not paying cash for the fuel wood, the forest was nonetheless producing it and they were consuming it. When the project prevents use of the forest for fuel wood with-out providing an alternative domestic energy source with lower GHG emissions, the demand for domestic energy is displaced to other forests. This constitutes leakage that should, at least in principal, be captured by the methods in Chapter 10. However, when displacement involves non-cash activities, analysts must use alternative approaches to estimate the relationship between supply and demand. As a rule of thumb, subsistence demands are usually relatively inelastic.

If activity shifting arises from a shift of labor and the shift occurs in a market economy, the methods in Chapter 10 account for this effect. For example, consider a situation where a logger is laid off because a project stops a logging enterprise that otherwise would have provided employment. If that logger finds another logging job with a different company, presumably that company is either expanding its logging or replacing another logger who stopped working. If the company is expanding its logging, the methods in Chapter 10 should account for that. If, on the other hand, the laid-off logger replaces another worker who has left the workforce, the job shift is not increasing logging and there is no leakage.

.

Appendix 21

Land-Management Projects and Changes in Demand

Terrestrial GHG mitigation projects seldom change demand for a product. Instead, such projects usually either increase or decrease product use.

Consider the case of increased product use. Because emissions generally result from the production or use of goods or services, production increases generally cause emission increases, not reductions. Obviously, activities that increase net emissions cannot be used to create offsets.

Some sequestration projects may cause future increases in product use. For example, projects that plant trees could increase the timber supply when the trees are available for cutting. In theory, this extra supply could reduce the price of wood, thus causing an increase in logging and negating some or all the sequestration achieved by the project. It is uncertain whether these effects in the distant future will ever occur. Technologies change over time, including those for making products now made of wood. Changes in tastes, wealth, or land-use rules could increase or decrease the use of wood. Because of uncertainty about future demand for wood, the methods presented here do not attempt to estimate these potential effects.

Consider another variation on increasing the use of a product: substituting a lower-emission product for a higher-emission product. For example, suppose that a project promotes the use of joists made of engineered wood in buildings, replacing the use of solid-wood joists. The engineered joists use less wood, allowing construction of the same number of square feet of new buildings with a smaller harvest of wood. The switch in joist technologies could decrease both the harvest and emissions from the harvest. The question is whether this decrease in emissions counts as offsets.

For the decrease in emissions to count, it must be direct to the project. If the maker of the engineered joists merely buys materials on the commodity wood market, the resulting harvest reduction will be indirect to the project, and joist makers and house builders cannot claim the emissions benefits as offsets. However, if a joist maker strikes a deal with a forest landowner to preserve harvestable timber equal to the wood not used, that timber can count as offsets, and the joist maker and the landowner owner can split any revenues.

Now consider the case when a project decreases demand for a product, such as by persuading people building new houses to create smaller rooms, thus providing new housing for the same number of people using less wood. This reduction in wood use reduces timber harvest and avoids some emissions of forest carbon. But as with the example of wood-use efficiency, the emission reduction is indirect to the people building houses. The decreased use of wood produces an emissions benefit, but the project does not own it and thus cannot claim it as offsets. To make the emissions benefits direct to the project, house builders would have to sign an agreement with a forest landowner to preserve some harvestable timber.

Appendix 22

Addressing Leakage from Forestation Projects

Most forestation projects that aim to sequester carbon exert a small impact on timber markets. Such markets can easily adapt to small changes in supply, so most forestation projects have high rates of leakage. For example, consider a moderately large project that removes 1,000 acres of 50-year-old coastal Douglas fir from the timber market. If harvested, this forest would have yielded about 10 million cubic feet of timber.

In 2002, the U.S. harvest of hardwood and softwood combined totaled more than 467 million cubic meters. (Howard 2004). To replace the lost 10 million cubic feet of timber, each U.S. supplier would have to begin production just 6.6 minutes earlier each year. In reality, a few suppliers would probably move up harvests by days or weeks to make up the difference. The point is that even a project of significant size is small relative to market flow and does not cause a significant disruption. This analysis shows how easily other suppliers can compensate for the withdrawal of one supplier from the market.

However, developers of forest projects can avoid leakage by maintaining the supply of final products. For example, a project developer could establish an intensively managed timber plantation to replace the supply lost from conserving existing forest. As long as carbon stock is not declining on the parcels where harvesting continues, no leakage would occur.

Project developers can also avoid leakage if the land use they displace is declining overall. For example, if a project reforests land previously used as pasture, and total pasture use is declining more quickly than offset projects are reforesting land, the project would not displace demand for pasture. (However, the baseline may have to take such changes into account—see Chapter 5.)

Leakage can rise or fall during forest-preservation projects. Consider a project that preserves a parcel of forest and records 100 percent leakage in the early years. If developers eventually convert all nonpreserved forestlands into buildings, roads, and parking lots, the remaining forest would avoid leakage. Because all nonpreserved forest is destroyed, and the only remaining forest is preserved as part of a sequestration project, the preserved forest actually does reduce emissions. However, only after most forestland has been developed does any reversal of leakage occur. If analysts develop methods for quantifying this reversal, project developers could count more sequestered carbon as offsets.

Forestation projects can produce leakage even if they *increase* the timber supply. That is because, besides displacing prior uses such as agriculture, such projects may spur other land managers to avoid some timber production because those managers expect timber from the projects to enter the market. However, studies have shown that this type of leakage is small. The approach to calculating displacement from expanded forestation is the same as that for other displacement, but with some added twists. If the project involves legal restrictions that prevent harvesting, it could displace tree planting for conservation.

If a project preserves young forests, no leakage would occur until those forests reached typical harvest age. However, the long delay between planting and harvest may make calculating displacement based on the price elasticities of supply and demand impractical. For example, projects that entail planting trees could increase the timber supply when the trees are available for cutting. The expanding supply could reduce the price of wood, prompting other landowners to raise production, negating some or all of the sequestration achieved by the project. However, changes in the techniques for making wood products, consumer tastes, wealth, and land-use rules could increase or decrease the use of wood. Because future demand for wood is uncertain, project analysts should not try to estimate future impacts.

Appendix 23

Using Regression Analysis to Calculate Elasticity

Analysts should use regression to calculate elasticity only if they correct for several factors. For example, they can adjust for inflation by including the consumer price index as an independent variable in the regression. Using a log-linear equation form often adjusts for nonlinear information. Serial correlation should be tested and corrected if present. One method for correcting for serial correlation is the Cochrane-Orcutt procedure. Analysts should impose a restriction that the function is homogeneous of degree zero in prices and nominal income (Attfield 1985). If analysts cannot meet the conditions for regression, they can use simultaneous equations to calculate elasticities.

For example, Kim (2004) finds the demand elasticity for a project that reduces rice acreage by regressing total U.S. rice consumption (Q_c) on rice price (P_c), the consumer price index (CPI), and total rice expenditures (EXP) using a log-linear functional form. He also uses a Cochrane-Orcutt procedure to correct for serial correlation and imposes a restriction that the function is homogeneous of degree zero in prices and nominal income. The results are

$$Ln\ Q_c = 0.9064 - 0.9139\ Ln\ P_c - 0.1672\ Ln\ CPI +$$
$$1.0811\ Ln\ EXP$$
$$(4.210)\quad (-13.610)\quad\quad (-4.788)\quad\quad (20.558)$$
$$R\text{-}Square = 0.924 \quad\quad DW = 2.00$$

Equation A23.1

where Ln is the symbol for the natural logarithm, the numbers in parentheses are t statistics, R-$Square$ is a goodness-of-fit indicator, and DW is the Durbin Watson statistic, which tests for the presence of serial correlation.

Using this equation, Kim finds that the demand elasticity (E) is −0.9139.

Appendix 24

Guidelines for Auditing Greenhouse Gases

The EPA's Emission Inventory Improvement Program (EIIP) lists five types of audits designed to address the quality of data companies report (see Table A.19). The EIIP includes specific instructions as well as general principles for conducting these audits, even providing sample checklists and auditors' reports. Although some of the more detailed suggestions do not apply to GHG inventories, the examples make excellent starting points for developing a GHG registry and accounting and verification protocols.

The U.S. Environmental Protection Agency's Greenhouse Gas Inventory Quality Assurance/Quality Control and Uncertainty Management Plan applies many of the principles outlined by the EIIP to GHG inventories, and it can also be an excellent foundation for defining verification tasks. Other guidelines for auditing emissions inventories range from broad principles (the Global Reporting Initiative's Sustainability Reporting Guidelines) to step-by-step checklists (the California Climate Action Registry) (see Table A.20).

Table A.19 Audit Types, as Classified by the EIIP

Audit type	Objective
Management systems	Determine the appropriateness of the management and supervision of inventory-development activities and training of inventory developers.
Technical systems	Determine the technical soundness, effectiveness, and efficiency of the procedures used to gather data and calculate emission results.
Performance evaluation	Determine whether the equipment used to collect quantification data operates within acceptable limits.
Data/report	Determine whether the results reported accurately reflect the emission results calculated and recorded in the supportive data.
Data quality	Determine the accuracy and completeness of the data used to develop the emission results.

Table A.20 Guidance on Verifying Emissions Inventories

	GHG specific?	Assumed inventory purpose	Notes
California Climate Action Registry	Y	Baseline protection	Step-by-step, publicly posted protocol.
This volume	Y	Create tradable offsets	Specific to offsets generated from terrestrial GHG sinks.
U.S. EPA Emission Inventory Improvement Program	N	Regulatory compliance	Thorough treatment of audit principles, including the Data Attribute Rating System for evaluating data quality.
U.S. Greenhouse Gas Inventory Quality Assurance/Quality Control and Uncertainty Management Plan	Y	Multiple	Most thorough treatment of quality-assurance/quality-control principles applied to GHG inventories.
World Business Council for Sustainable Development and World Resources Institute (WBCSD/WRI GHG) Protocol	Y	Corporate accountability	General principles only; no procedures.
GRI Sustainability Reporting Guidelines	N	Corporate accountability	General principles only; no procedures. Appropriate for corporate self-audits only.
International Standards Organization 14001	N	Corporate accountability; regulatory compliance	Governs auditing of corporate environmental-management systems only.
Kyoto/CDM	Y	Regulatory compliance	Marrakech Accords provide general guidelines for auditing offset projects. CDM Executive Board may release specific guidelines in the future. One of the few sources of guidance on validating offsets.
Kyoto/inventory review	Y	Regulatory compliance	General guidelines for auditing national GHG inventories from the eighth Conference of the Parties. Some principles also apply to offset projects.
Environmental Resources Trust: Corporate Greenhouse Gas Verification Guideline	Y	Multiple	Designed for corporation-wide audits.

Appendix 25

Verifying and Registering Offsets under the Kyoto Protocol

The Kyoto Protocol's Clean Development Mechanism (CDM) is developing an extensive by-the-project system for validating and verifying greenhouse offsets. The CDM offers countries committed to reducing their greenhouse gas emissions (those with emissions caps and listed in Annex B of the Kyoto Protocol) the opportunity to sponsor GHG-reduction projects in developing countries that do not have emissions caps. Rigorous validation and verification of these projects is important because project developers must define their boundaries and calculate baselines, additionality, and leakage.

To be fully vetted in the Kyoto CDM system, projects must follow several steps (see Figure A.14). The system's rigor lies in the fact that each project passes through the hands of two third parties, known as Designated Operational Entities (DOEs). The first DOE validates the project. Besides certifying that the design is competent, validation ensures that the CDM registry will accept verified and certified reductions reported later.

The second DOE performs verification and certification: "Verification is the periodic independent review and *ex post* determination by the designated operational entity of the monitored reductions in anthropogenic emissions by sources of greenhouse gases that have occurred as a result of a registered CDM project activity during the verification period. Certification is the written assurance by the designated operational entity that, during a specified time period, a project activity achieved the reductions in anthropogenic emissions

by sources of greenhouse gases as verified" (United Nations 2002).

The Kyoto Protocol's Marrakesh Accords allow a project to request and receive approval to use the same DOE for validating and verifying offsets, but note that this should be the exception. The CDM Executive Board does not specify a schedule for verification activities. It is the DOE's responsibility to perform verification appropriately so it supports certification for the crediting period defined in the project design. This period can be up to seven years long and renewed twice or 10 years long and nonrenewable.

Parties apply to the CDM Executive Board to become accredited DOEs. The board's Accreditation Panel appoints a team to review the application, conduct an assessment of the entity applying to become a DOE, and witness an applicant perform its first validation or verification. If the panel is satisfied with the results, it recommends that the Executive Board accept the entity as a DOE for the type of project for which the entity has demonstrated competence in validating or verifying.

The Kyoto Protocol specifies a second mechanism besides the CDM, known as Joint Implementation (JI), for creating GHG offset projects within countries that have agreed to cap their emissions under the protocol. Unlike the CDM registry, the protocol does not directly mandate a central JI registry, so independent registries must set validation, verification, and accreditation standards for JI projects. The Dutch ERUPT system is leading the way, with a validation-verification setup similar to the CDM mechanism.

Those wishing to become accredited third parties (DOEs) under ERUPT apply to the Dutch Board for Accreditation. As under the CDM scheme, applicants are subjected to a desk review, site inspection, and witness assessment. The Dutch board may streamline the process for certifying a DOE already approved by the CDM process.

The Kyoto Registry System

The Kyoto system is remarkable for supporting multiple registries and a hybrid of closed and open markets. Kyoto registries track four types of emissions allowances. One type is permission to generate one metric ton of CO_2e of anthropogenic GHGs at any time during the 2008–2012 compliance period. At the end of 2012, a true-up occurs, and the system retires allowances

Figure A.14 The process followed by an offset in the Kyoto Clean Development Mechanism (CDM) process. The CDM offers countries committed to reducing their greenhouse gas emissions—those that have emissions caps and are listed in Annex B of the Kyoto Protocol—the opportunity to sponsor GHG-reduction projects in developing countries, which do not have emissions caps, and in so doing to get credit for emissions reductions as if they occurred in their country. To be fully vetted in the Kyoto CDM system, projects must follow the steps illustrated here.

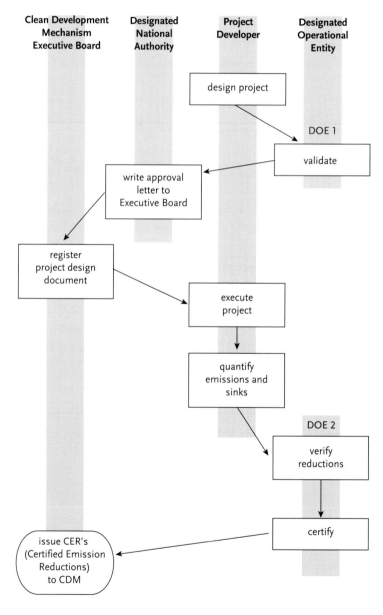

equal to each country's emissions during the compliance period. The protocol defines basic allowances assigned by Article 3 (known as assigned amount units [AAUs]) as five times a designated fraction of each country's 1990 emissions. The combined AAUs form the 2008–2012 cap.

Removal units (RMUs) are additional allowances, each representing one metric ton of CO_2e of anthropogenic GHG that is offset by land management. RMUs allow each country to inflate its portion of the gross emissions cap, limited by that country's capacity to absorb an equivalent amount of carbon in forests. Although RMUs are created by the Kyoto Protocol, they are not now traded. Instead, countries with emission caps trade sink amounts by turning them into national allowance units (AAUs) or emission reduction units (ERUs).

JI projects convert AAUs to ERUs, allowing emission reductions to move from one country with an emissions cap to another. Because ERUs are converted AAUs, they do not affect the market cap. ERUs give developing countries confidence that the allowances they sell are backed by a true reduction in national emissions.

Certified emission reductions (CERs) are allowances generated outside the market cap by non–Annex B countries and sold to Annex B countries inside the cap. CERs are an open-market commodity sold into a closed market, and they have the effect of increasing the total number of tradable units without raising global net emissions. Because RMUs and CERs can expand the closed-market cap defined by the quantity of AAUs, the Kyoto system is a hybrid of closed and open markets.

The system uses three types of registries to manage this complex arrangement: national registries, CDM registries, and the transaction log of the U.N. Framework Convention on Climate Change (UNFCCC). Each Annex B country operates a national registry that tracks the emissions allowances assigned to, generated by, and purchased by that country. A single CDM registry operated by the CDM Executive Board holds and trades CERs generated by non–Annex B countries, which do not maintain their own registries. Each non–Annex B country that sells CERs must write a letter of approval authorizing the transfer of these tons to other countries. The designated national authority that writes these letters does not have to track the amounts or ensure that they meet any quality standard. The UNFCCC secretariat maintains the transaction log—a registry of trades among national registries and between national registries and the CDM registry. The transaction log determines that

– The transferred units are not retired or canceled.
– The units do not exist in more than one account.
– The units were not improperly issued.
– The requesting parties are authorized to request the transfer.
– The modified content of each account does not violate the protocol.

Appendix 26

Choosing a Registry

Offsets gain market value when they are vetted and made accessible through registration. But which registry provides offsets with the most legitimacy and therefore the highest value? The answer depends on whether sellers are seeking financial rewards or recognition. (See Figure A.15 for a general roadmap for choosing a registry, beginning with the fundamental question, "Do I want to sell?")

Regardless of the seller's purpose in registering offsets, selecting a registry *before* initiating a project is essential. There are two reasons for this. The first is that some registries have specific requirements for validating projects during the design phase. The second is that a registry that will accept the proposed offsets may not exist, depending on the nature of the project and its geographic location.

Registration under the Kyoto System

If a project developer wishes to create fungible (that is, fully tradable) offsets, then the registries supporting the Kyoto Protocol are the most thoroughly considered, often used as a model by other registries in the making (see Appendix 25). If a project's host country has ratified the Kyoto Protocol, registration under Kyoto is the most desirable destination for a land-management offset.

How a project fits into the Kyoto system depends entirely on its host country. Annex B countries have been assigned emissions caps under the protocol (see Table A.21). If the project is in a non–Annex B country, then it falls under the purview of the Clean Development Mechanism (CDM). Many types of energy and industrial emissions-reduction projects qualify for CDM registration. However, among land-management efforts, the system now recognizes only reforestation projects. To register a project under the CDM, developers must prepare a design document for the CDM Executive Board before beginning the project.

If the project is in an Annex B country, then the country must be an "economy in transition." Such countries are permitted to sell offsets from Joint Implementation (JI) projects to countries with stronger economies (see Appendix 25).

Countries with emissions caps may include sequestration resulting from changes in agricultural soil, land use, and forestry in their national inventories of emissions. These sinks could create emission reduction units (ERUs), which could be traded under the JI program. No systems yet exist for creating ERUs from changing land-management activities. Project developers would be wise to open negotiations with their federal government early to help create a market for land-management offsets.

Registering Offsets for Financial Gain outside the Kyoto System

Several regions around the world are establishing systems for trading GHG allowances or credits outside

Figure A.15 Roadmap for choosing an offset registry. For an offset to have market value it must be registered with an independent agency or registry. The type of registry to choose depends upon a variety of factors, including the nature and the goals of the project.

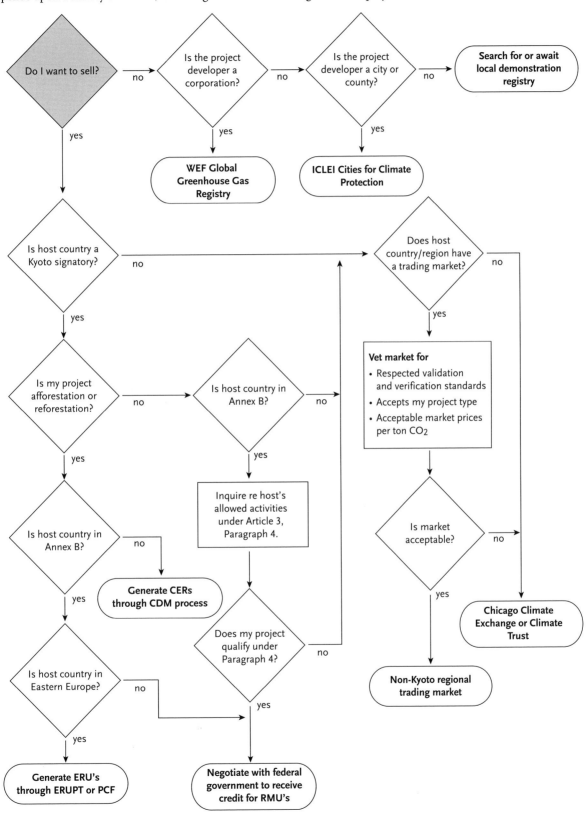

Table A.21 Countries That Have Ratified the Kyoto Protocol

Antigua and Barbuda	Georgia	**New Zealand**
Argentina	**Germany**	Nicaragua
Armenia	Ghana	Niue
Austria	**Greece**	**Norway**
Azerbaijan	Grenada	Paulau
Bahamas	Guatemala	Panama
Bangladesh	Guinea	Papua New Guinea
Barbados	Guyana	Paraguay
Belgium	Honduras	Peru
Belize	**Hungary***	Philippines
Benin	**Iceland**	**Poland***
Bhutan	India	**Portugal**
Bolivia	**Ireland**	**Romania***
Botswana	Israel	**Russian**
Brazil	**Italy**	**Federation***
Bulgaria*	Jamaica	Saint Lucia
Burundi	**Japan**	Samoa
Cambodia	Jordan	Senegal
Cameroon	Kiribati	Seychelles
Canada	Kyrgyzstan	**Slovakia***
Chile	Laos	**Slovenia***
China	**Latvia***	Solomon Islands
Colombia	Lesotho	South Africa
Cook Islands	Liberia	South Korea
Costa Rica	**Lithuania***	**Spain**
Croatia	**Luxembourg**	Sri Lanka
Cuba	Madagascar	**Sweden**
Cyprus	Malawi	**Switzerland**
Czech Republic*	Malaysia	Tanzania
Denmark	Maldives	Thailand
Djibouti	Mali	Trinidad and
Dominican	Malta	Tobago
Republic	Marshall Islands	Tunisia
Ecuador	Mauritius	Turkmenistan
El Salvador	Mexico	Tuvalu
Equatorial Guinea	Micronesia	Uganda
Estonia*	Moldova	**Ukraine***
E.C.	Mongolia	**United Kingdom**
Fiji	Morocco	Uruguay
Finland	Myanmar	Uzbekistan
France	Namibia	Vanuatu
Gambia	Nauru	Vietnam
	Netherlands	

Notes: Annex B countries are listed in boldface; Annex B countries that may host Joint Implementation projects are indicated by an asterisk.

the Kyoto Protocol. The European and U.K. emissions-trading systems are examples, but they do not register land-management offsets. The New South Wales GHG Abatement Scheme in Australia is the only system now accepting such offsets—from afforestation and deforestation only. Land-management offsets created outside New South Wales may qualify for that market under limited conditions. Other emissions-trading systems are in early stages of development in Canada, the northeastern U.S., and elsewhere, and they may eventually allow sales of offsets.

If project developers cannot find a regional trading market, then a private market is the only remaining option. One such market, the Chicago Climate Exchange, is operating demonstration trades, and it accepts some types of land-management offsets created in North America and Brazil. Another market, Climate Trust, acts more as a broker with high standards than as a registry, but it is open to creative approaches to GHG reduction and can be an effective conduit for offset sales. Prices for offsets registered with the Climate Trust have ranged from $2 to $10 per metric ton of CO_2e—a testament to the rigor of the registry's standards.

Registering Offsets for Recognition

Some project developers may prefer to register offsets without financial gain because they want to

- Track their gross annual emissions.
- Achieve voluntary cuts in emissions.
- Reap public-relations benefit from their efforts to create offsets.
- Ensure recognition of early action on greenhouse emissions because they anticipate future regulation.

Two worldwide programs exist for this purpose. The World Economic Forum's Global GHG Registry records companies' greenhouse emissions and reductions. The International Council for Local Environmental Initiatives administers Cities for Climate Protection (CCP) for cities, counties, and other small governments. CCP is not a registry, but it does provide standards and tools for consistent GHG accounting.

Table A.22 GHG Recognition Programs That Accept Land-Management Offsets

Organization	Domain	Can offsets originate outside of domain?	Types of offsets allowed	Entity-wide reporting required	Web address	Accept sequestration?
U.S. DOE 1605(b)	U.S.	No	6 Kyoto GHGs + others	No	www.eia.doe.gov/oiaf/1605/frntvrgg.html	Yes
U.S. EPA Climate Leaders	Global	n/a	8 GHGs (still finalizing methodology)	Yes	www.epa.gov/climateleaders/	Not final; probably yes
State of Wisconsin	Wisconsin	No	CO_2, methane, hydrofluorocarbons, CO, VOC, mercury, lead	No	www.dnr.state.wi.us	Yes
State of New Jersey	New Jersey	Possibly	CO_2 and methane	No	www.nj.gov/dep/aqm/	Not yet determined
California Climate Action Registry (CCAR)	California	Yes, U.S.	8 GHGs	Yes	www.climateregistry.org/	Not yet determined
Chicago Climate Exchange	North America	Only if from Brazil	Soil, forest, methane, CO_2, 6 GHGs	Yes	www.chicagoclimatex.com/	Yes
Global Greenhouse Gas Register	Global	n/a	6 Kyoto GHGs, but reported separately from emissions totals	Yes	www.weforum.org/ghg	Yes, but report separately
Climate Neutral Network	Global	n/a	8 GHGs	No	www.climateneutral.com/	Yes, some types
Environmental Defense	Global	n/a	6 Kyoto GHGs (any if surplus)	Yes	www.pca-online.org	Yes
Greenhouse Gas Protocol (WRI/WBCSD)	Global	n/a	8 GHGs	Yes	www.ghgprotocol.org	Not yet determined
Canada's Voluntary Challenge	Canada	Yes	6 Kyoto GHGs	No	www.vcr-mvr.ca	Yes
Australia GHG Challenge	Australia	No	6 Kyoto GHGs (not officially accepting offsets)	Yes	www.greenhouse.gov.au/challenge/	Not yet operational
EcoGESte (Quebec)	Quebec	Possibly	6 Kyoto GHGs	No	www.ecogeste.gouv.qc.ca	Not yet determined
Environmental Resources Trust	U.S.	Yes	CO_2, methane, NOx, SOx	Yes	www.ert.net	Yes

Notes: U.S. DOE = U.S. Department of Energy; U.S. EPA = U.S. Environmental Protection Agency; VOC = volatile organic carbon; NOx = nitrous oxides; SOx = sulfur oxides.

If neither program is appropriate, several countries also offer programs through which parties can set voluntary GHG reduction targets. Some of these programs include a rudimentary registry through which participants can report their annual GHG inventories and the results of offset projects. In the United States, a few states, including New Hampshire and Wisconsin, have established demonstration GHG registries that accept offsets. Table A.22 provides a sampling of recognition programs throughout the world.

Appendix 27

Sample Field Protocol: Establishing Plots and Measuring Biomass in a Forestry Project

This appendix provides a step-by-step guide to establishing sampling plots and measuring biomass for a forestry project.

Equipment List

Vicinity map

Site map marked with plot-center locations and coordinates

Sampling protocol and tree-identification guide

Data sheets, data sheet holder, pencils

Plot-layout graphic

Slope-correction table

Calculator

GPS receiver and batteries

Compass

Rebar & PVC tube, 1 each per plot

Hammer or small sledge

Rangefinder (with or without reflector and reflector holder) (optional)

25 m tape

Wire flags

15 cm ruler

Diameter tape

Clinometer (device for measuring the angle of a line of sight above or below horizontal—used to measure tree height)

Optical dendrometer (instrument for measuring tree height and girth) (optional)

Calipers with 1, 3, 5, and 10 cm fixed gaps (saplings and wood debris)

Adjustable caliper, capacity up to the maximum tree diameter

Sampling frame for measuring litter, 0.5m × 0.5m, that can be disassembled

Pruning shears

Permanent marker

Ziploc bags, gallon size, 1 each per plot

Compactor bags, 3

Spring scales, 100 gm, 1 kg, 10 kg

Machete (optional)

Bug repellent, poison ivy block, raingear (optional)

Personal safety and health supplies: water, food, first aid kit, radio or phone (optional)

Establishing Plots

Note the start time when leaving the vehicle. Using the vicinity map, site map, plot-center coordinates, and GPS, go to the plot center. Note any permanent landmarks useful for future relocation of plot. Ensure that compass readings are made away from significant magnetic objects, such as the vehicle, rebar, and steel tools. Note the time at which the plot center is located and reached.

Pound the rebar into the ground at the plot center, leaving a few centimeters above the ground to use as a stake for tapes used to measure distance from the plot center. Hook the tape on the rebar, and place the PVC

Table A.23 Slope Correction

Slope (percent)	Slope distance				
	0.1 ha plot	0.01 ha plot	0.002 ha plot	0.005 ha plot	0.001 ha plot
0	17.84	5.64	2.52	3.99	1.78
5	17.86	5.65	2.53	3.99	1.79
10	17.93	5.67	2.54	4.01	1.79
15	18.04	5.71	2.55	4.03	1.80
20	18.19	5.75	2.57	4.07	1.82
25	18.39	5.82	2.60	4.11	1.84
30	18.63	5.89	2.63	4.17	1.86
35	18.90	5.98	2.67	4.23	1.89
40	19.22	6.08	2.72	4.30	1.92
45	19.56	6.19	2.77	4.37	1.96
50	19.95	6.31	2.82	4.46	1.99
55	20.36	6.44	2.88	4.55	2.04
60	20.81	6.58	2.94	4.65	2.08
65	21.28	6.73	3.01	4.76	2.13
70	21.78	6.89	3.08	4.87	2.18
75	22.30	7.05	3.15	4.99	2.23
80	22.85	7.23	3.23	5.11	2.28
85	23.42	7.40	3.31	5.24	2.34
90	24.00	7.59	3.39	5.37	2.40
95	24.61	7.78	3.48	5.50	2.46
100	25.23	7.98	3.57	5.64	2.52

pipe over the end of the rebar. Attach the GPS receiver to the top of the PVC pipe, and record the GPS reading of the actual plot center location. The GPS receiver is easily attached to the PVC by tying a 1-centimeter-diameter stick to the receiver and sliding the end of the stick into the top of the PVC pipe.

Measuring Woody Debris

Start by measuring medium debris, before the plot is trampled. Establish a north-south transect and an east-west transect, both passing through plot center. Make transects 50 meters long, extending 25 meters from the plot center in each of the four cardinal directions. For each plot, use the same transects for all woody debris, litter, and shrub and sapling measurements.

At the plot center, use a compass to sight south, and note the true due-south line relative to terrain features. Starting from the plot center and moving south along the transect, stretch a tape. Along the first 5 meters of the transect heading north from the plot center, record the diameter of each piece of debris greater than or equal to 1 centimeter in diameter but less than 10 centimeters in diameter. For each piece, record the decay class—hard or soft. Count suspended pieces that cross the transect, as long as the roots of the plant are not still attached to the piece and in the ground. Record decay classes as H for hard and S for soft. Hard debris is debris that would break with a cracking sound or splinter and that would give a solid or ringing sound if struck with a machete. Soft debris would break quietly and would typically break close to a plane perpendicular to the growth axis. Soft debris would also give a soft or squishing sound if struck with a machete.

Complete the location of the south transect to its full 25-meter length. While measuring the woody debris transects it is efficient to also mark the boundaries of the tree plots. Placing a flag at 5.64 meters and 17.84 meters from the plot center marks the edges of the medium and large tree plots and may speed determination of whether individual trees are inside or outside the plot. If the slope is greater than 10 percent, correct for it in measuring these distances. Sight back north, and establish the coarse-woody-debris transect extending north from the plot center. Make a compass sighting from plot center to locate the east or west transect. Sight back along it in order to locate the other east-west coarse-woody-debris transect. As with the south and north transects, placing flags at 5.64 meters and 17.84 meters from the plot center may speed determination of whether individual trees are inside or outside the plot. As these transects are established, measure and record information on coarse debris while the tape is still stretched along the transect line, before establishing the next line. (Coarse woody debris is defined as dead woody material where the growth axis is more than 45 degrees from vertical or dead woody material of any orientation where the piece is not attached to its roots or the roots are not in the ground.) Record the diameter and decay class (1 through 5—see below) of each piece of debris that crosses the transect and is 10 centimeters or greater in diameter. If the species can be identified, record the species.

If working as a multiperson team, first establish and flag the coarse-debris transects, and then take measurements. If there is uncertainty about whether a piece crosses the transect, the person collecting data can hold a finger or a flag above the midline axis of the piece, and the person recording data at plot center can sight to the flags farther out the transect and judge whether the piece crosses it. Be careful not to disturb litter in the vicinity of the litter plot.

Measuring Litter and the Organic Layer

Fine woody debris—less than 1 centimeter in diameter—is measured with litter. Go to the flag placed 5.64 meters south of the plot center: this is the northeast corner of the litter plot. The litter plot is 0.5 by 0.5 meters, with its edge aligned with the debris transect. Place the frame for the litter quadrat on the ground. If needed, take apart the frame to place it around the stems of shrubs and below the canopy of shrubs, and reassemble it in the desired position. Hold one edge or corner of the frame up, as needed, so that the frame is horizontal. If the slope is too steep to accurately locate the subplot edges when the frame is held horizontally, lay the frame on the ground surface. If the frame is placed on sloping ground, rotate the frame clockwise around the reference-corner point until one pair of opposing frame edges is horizontal across the slope. No more than 90 degrees of rotation is needed. Record the slope of the sloping pair of frame edges. (Make sure slope is recorded in percent, not degrees.) Record the slope of the frame when it is used to determine what litter to collect. If the frame is held horizontally on sloping ground, record the slope as zero.

Use the pruning shears to cut the litter along the line below the inside edge of the sampling frame. Put the litter and duff in a bag of sufficient size (typically a compactor bag, but perhaps a gallon Ziploc), and weigh it on a tared scale.[1] Do not include live vegetation or live moss. Record the weight of the litter. On the bag, write the site name or code and the plot number. Tare the appropriate scale for the bag. Take a representative subsample of not more than about 1 liter in volume and not less than about 100 grams, put it in the bag, and weigh it on the tared scale. Write the field weight on the bag and on the data sheet. Take the subsample back to the lab for analysis.

In the quadrant where litter was removed, use the trowel to cut vertically into the soil. Pull the trowel to one side, exposing a vertical face of soil. Use visual inspection and feel for grittiness when rubbing a small amount of soil between fingers. An organic layer feels smooth or creamy when smeared between the fingers, perhaps with soft organic lumps. Organic material is dark brown or black. Mineral soil feels gritty or hard, except for pure clay. Identify the depth of any organic layer. Record the depth. Repeat this depth measurement for the required number of measurements, spreading the measurements evenly around the quadrat area.

Measuring Saplings and Shrubs

Go to the flag 5.64 meters north of the plot center on the woody-debris transect. Place the 0.5-by-0.5-meter quadrat frame over the flag, with the flag forming the northeast corner of the quadrat, or use a tape to measure the quadrat. If using the frame, hold up the frame, if necessary, to correct for slope. Count the number of live woody stems originating from within the quadrat that are up to 1 centimeter in diameter at the base. Record the count, median diameter, height, and species providing the largest percentage of cover within the quadrat.

Establish a plot with a radius of 1.78 meters, centered on the flag. Slope-correct all horizontal measurements if the slope is greater than 10 percent. Count all live woody stems originating within the plot that are 1 to 3 centimeters in diameter at the base. Record the count, the median diameter at the base, the height, and the species with the largest percentage of cover within the quadrat. Do the same for all live plants that are 3 to 5 centimeters in diameter at the base, and for all live plants that are greater than 5 centimeters in diameter at the base but less than 5 centimeters in diameter at breast height (dbh).

Measuring Trees

Establish a plot with a 2.52-meter radius, centered on the plot center. Slope-correct all horizontal measurements if the slope is greater than 10 percent. Working clockwise from map north, for each stem that is at least 5 centimeters dbh and that has its center at ground level within the plot, record: species code, diameter, height, top diameter, and vigor/decay class. Measure the top diameter of broken trees using the optical dendrometer. Include all live stems, snags, and stumps of the size class in each subplot. Snags and stumps are defined as dead plants with woody stems still attached to their roots, with the roots mostly in the ground, and with a growth axis within 45 degrees of vertical. Stumps are very short snags. When stumps are less than 1.37 meters tall, record the actual top diameter.

Establish a plot with a 5.46-meter radius centered on the plot center, slope-corrected as necessary. Working clockwise from map north, for each stem originating within the plot that is at least 15 centimeters dbh, record: species code, diameter, height, top diameter, and vigor/decay class. Include all live stems, snags, and stumps.

Establish a plot with a 17.84-meter radius centered on plot center, slope-corrected as necessary. Working clockwise from map north, for each stem originating within the plot that is at least 30 centimeters dbh, record: species code, diameter, height, top diameter, and vigor/decay class. Include all live stems, snags, and stumps.[2]

It may be more efficient to record all live trees of a given radius on a plot and then to record snags, especially if the plot contains numerous stumps or other short snags.

Wrapping Up

Roll all the tapes. Pound the rebar until it is flush with the ground or slightly below the surface. Write the plot number near the top of a 1-meter-long piece of PVC pipe using permanent marker. Push the PVC pipe a few inches over the top of the rebar so it stands and marks the plot center. Pick up all the flags and stakes other than the plot-center marker. Record the time of leaving the plot. It is good practice to look over the plot for forgotten tools; this takes only a few moments and can avoid frustration and hours of time spent looking for and replacing lost tools.

Codes for Recording Species on Data Sheets

The code for an individual species is the first two letters of the genus followed by the first two letters of the species. If two species would have the same code, add a number after the species code, such as ACSA2. When plants are identified to genus only, use the first four to six characters of the genus as the code. Species codes must be tailored to each project area; the codes below illustrate the format.

Shrub Species

ALNUS	Alnus spp.	Alder species
CORNUS	Cornus spp.	Dogwood
SAEX	Salix exigua	Narrowleaf willow, all variants
SAMB	Sambucus spp.	Elder

Tree Species

ABBA	*Abies balsamea*	Balsam fir
ACNE2	*Acer negundo*	Box elder
ACSA2	*Acer saccharinum* L.	Maple, silver
ACSA3	*Acer saccharum*	Maple, sugar, or hard maple
AEGL	*Aesculus glabra*	Ohio buckeye
BELU	*Betula lutea Michx.* F.	Birch, yellow
BENI	*Betual nigra*	Birch, river
CACO15	*Carya cordiformis* (Wang.)	Hickory, bitternut, or yellowbud
CAOV2	*Carya ovata* (P. Mill.)	Hickory, shagbark
CEOC	*Celtis occidentalis* L.	Hackberry
FRAM2	*Fraxinus Americana* L.	Ash, white
QUERC	*Quercus species*	Oak species
QUMA2	*Quercus macrocarpa Michx.*	Oak, bur
QURU	*Quercus rubra* L.	Oak, northern red
QUVE	*Quercus velutina. Lam*	Oak, black

ROPS	*Robinia pseudo-acacia* L.	Locust, black
SANI	*Salix nigra Marsh.*	Willow, black
TIAM	*Tilia Americana* L.	Basswood
TSCA	*Tsuga canadensis* (L.) Carr.	Hemlock, eastern
ULAM	*Ulmus americana* L.	Elm, American
ULRU	*Ulmus rubra Muhl.*	Elm, slippery
UNKN		Unknown

Tree Vigor/Snag Decay Class

L1	good vigor	No apparent signs of distress (e.g., discolored leaves, paucity of leaves, conks, significant stem or root rot likely to cause falling).
L2	fair vigor	Some signs of distress apparent.
L3	poor vigor	Extreme distress apparent, imminent death likely.

D1 Dead, bark intact, fine branches present.
D2 Dead, hard snag, medium branches present, top may be missing, some bark likely missing.
D3 Dead, top and branches missing, substantial rot, hard core, may have some bark.
D4 Dead, soft snag.
D5 Dead, soft snag, rotted down to near stump, blob.

Decay Classes for Fine Medium Woody Debris (<10 cm diameter)

H Hard, breaks with snapping sound.
S Soft, breaks straight across grain, punky.

Decay Classes for Coarse Woody Debris (>10 cm diameter)

1. Hard, bark mostly intact, branches not rotted.
2. Hard but losing bark, fine branches rotted off.
3. Soft exterior, hard interior, not totally conforming to ground topography.
4. Soft, conforming to ground topography, partially buried.
5. Decayed to chunks or mush, substantially buried.

Sample Data Sheet—Biomass Measurement

BIOMASS SURVEY   Plot No.:_____   Time arrive at plot: _____
Surveyor(s):   Datum: WGS84 NAD27 NAD83   UTM zone/Lat: ___
Site name: Assigned Plot Center: _ _ _ _ _ _ _ _ mE _ _ _ _ _ _ _ _ mN
Date: Installed Plot Center: _ _ _ _ _ _ _ _ mE _ _ _ _ _ _ _ _ mN
Landmarks/Notes:

                                                     Time start sampling: _____
FINE DEBRIS _____ pieces 1.0–9.9cm dia, 5 m long transect south
LITTER Slope ___%   Weight: _____gm (0.5m × 0.5m) Subsample weight: _____gm
O-layer depths: ___ ___ ___ ___ ___ ___ ___ ___ ___ ___ ___ ___ ___ ___ cm

FINE & COARSE DEBRIS fine: <10 cm dia; 5 m transect south; coarse: four 25 m transects

Diameter	Decay Class	Species		Diameter	Decay Class	Species

SHRUB/SAPLING (5.64 m north) (1.78 m = 5'10")

DBA Class	Stem Count	Median dba (cm)	Median height (cm)	Species	Plot Size
0–1 cm					0.5m × 0.5m
1–3 cm					.001 = 1.78 m radius
3–5 cm					.001 = 1.78 m radius
>5 cm					.001 = 1.78 m radius

TREE & SNAG PLOTS

>=5.0 cm dbh: .002 ha plot radius 2.52 m = 8'3"

>=15.0 cm dbh .01 ha plot radius 5.64 m = 18'6"

>=30.0 cm dbh: .1 ha plot radius 17.84 m = 58'6"

Plot Size	Tree #	SPP	dbh (cm)	Height (m)	Top dia.	Vigor/ decay

Plot Size	Tree #	SPP	dbh (cm)	Height (m)	Top dia.	Vigor/ decay
			End	time		

Vigor: L1 = no apparent signs of distress (discolored leaves, paucity of leaves)

L2 = some signs of distress

L3 = extreme distress; imminent death likely

D1 = dead, fine branches present; CWD suspended

D2 = branches & bark generally present; CWD same

D3 = sapwood soft or gone; CWD same

D4 = soft, top gone; CWD soft & conforms to terrain

D5 = blob; CWD largely buried

Notes

Chapter 1: The Role of Landowners and Farmers

1. CO_2 emissions contribute most to global warming, but methane (CH_4), nitrous oxide (N_2O), and chlorofluorocarbon emissions also contribute. Of these last three, land-use practices affect only methane and nitrous oxide emissions, and thus are relevant here.

2. See Annex B of the Kyoto Protocol for a list of countries and the emission reductions they committed to during the negotiation of the Protocol. The U.S. and Australia have not agreed to implement their emissions reduction commitments.

3. See http://www.us-cap.org.

4. The United States pioneered the first cap-and-trade system designed to curb dangerous industrial emissions. In 1990, Congress amended the Clean Air Act to establish such a system to regulate releases of sulfur dioxide (SO_2), which is a precursor to acid rain. The program achieved 100 percent compliance at a fraction of projected costs. For a general primer on cap-and-trade systems, see: www.us-cap.org.

5. Offsets are sometimes called credits. The term *credits* is not used here because it may also to refer to emissions allowances. While offsets are typically referred to as *carbon* offsets, they include all reductions in greenhouse gases, not just CO_2. See Chapter 2 for how to assess the relative climate benefit of cuts in different greenhouse gases.

6. Additional information is available at: http://www.fightglobalwarming.com.

7. Dangerous climate change includes the melting of the major polar ice sheets, with an attendant rise in sea level of more than 20 feet and the loss of the Amazon rainforests.

8. House et al. 2002 provide a similar analysis from a global perspective.

9. For example, the Kyoto process recognizes only offsets generated from specified countries, using specified methods, and having passed specified reviews. It categorically excludes some of the methods for mitigating greenhouse gases recommended here.

Chapter 2: The Process of Creating Offsets

1. The definition of this quantity (CO_2 equivalent) is provided later in the chapter.

2. Emitters operating under a mandatory cap-and-trade system do not have to consider *additionality* when identifying the internal activities and practices that will be adopted to meet the emissions allowances or cap. That's because their emissions cap is their baseline, and any reductions in emissions represent a GHG gain. Of course, if a capped emitter were to purchase carbon offsets from a landowner to help meet its cap, those purchased offsets would indeed have to be additional.

3. Technically, a molecule of CO_2 emitted to the atmosphere generally leaves the atmosphere more quickly than a molecule of methane. Most CO_2 leaving the atmosphere is absorbed by oceans, however, which in turn emit more CO_2 as the concentration in the water rises. The release of a ton of CO_2 into the atmosphere thus sparks a chain of absorptions and emissions.

4. The small difference between the 1995 and 2001 reports reflect scientific improvements in the ability to calculate GWP over the six-year period. The Kyoto Protocol has not adopted the 2001 GWPs, so projects that generate offsets within the Kyoto system use the 1995 values.

Chapter 3: Land-Management Options

1. Unless otherwise noted, prices are in 2006 dollars.

2. Analysts estimate that between 1850 and 1990, one third of anthropogenic carbon emissions stemmed from vegetation and soils (see Houghton 1999).

3. Data are not available for the last few years. From the late-1980s to the mid-1990s, however, total U.S. forest carbon stock grew, primarily because of regrowth of forests on agricultural lands abandoned during the first half of the twentieth century.

4. Soil organic matter is heavily decomposed material that is relatively resistant to further decomposition, and is typically about 58 percent carbon by dry weight.

5. Crop breeders have focused on raising yields, sometimes by selecting plants that devote more resources to the crop and less to non-crop portions. Residue mass usually rises, however, even if it is a smaller proportion of total plant biomass.

6. U.S. Department of Agriculture, Farm Service Agency. See: http://www.fsa.usda.gov/pas/.

7. Unlike plowing, soil mixing by earthworms does not cause loss of soil carbon because the earthworms do not break up soil aggregates.

8. For more on land-management practices that can mitigate nitrous oxide and methane emissions, see Cicerone 1983, 1992; Wassmann, Papen, Rennenberg. 1993; Wassman, Wang, Shangguan, Xie, Shen, Papen, Rennenberg, Seiler, 1993; Denier van der Gon, 1993 and 1996; Zhou 1994; Neue 1994; Harrison 1995; Peoples 1995; Mosier 1996; Mosier (methane) 1998; and Mosier (nitrous oxide) 1998.

9. Denitrification, however, can lead to nitrous oxide emissions. Aerobic treatment of manure also produces substantial amounts of nitrate, which is a significant cause of water pollutants if not captured or converted to inert nitrogen.

10. The IPCC includes gathering dung for fuel as a waste-management practice that typically causes significant methane production because of incomplete combustion. An offset project that collects manure left in pastures for use as fuel would reduce nitrous oxide emissions, but higher methane emissions might offset this greenhouse benefit. This land-management change could also reduce stocks of soil carbon. The net impact of such a project would depend on the previous fuel and how it was burned. This manual does not include methods for calculating the greenhouse impact of switching to manure for fuel because that is not a land-management activity.

Chapter 4: Scoping the Costs and Benefits

1. The literature on performing such assessments is voluminous, and consultants are available to help.

Chapter 6: Carbon Sequestered in Forests

1. The word *reforestation* is used here in its ecological sense. In the business of forestry, the word *reforestation* refers to planting or natural regeneration to establish trees on a site that has been recently cleared by harvesting (especially clear cut harvesting) or some other disturbance. In forestry, the word *afforestation* is used to denote the reestablishment of forest on lands that have been in non-forest cover.

2. Recent scientific studies suggest that some aspects of a forest's life cycle may lead to climate warming instead of cooling (cf., Gibbard et al. 2005; Keppler et al. 2006). If these studies are confirmed by subsequent research, a plan for quantifying these warming effects would need to be incorporated into the measurement scheme (Olander 2006).

3. Developers may use less expensive methods than those recommended here to quantify the tons of CO_2e their project sequesters, but the former would provide much less reliable results, and thus could yield fewer tradable offsets.

4. If equations are available, it is possible that all field measurements will be of the aboveground attributes of trees, and belowground biomass may be inferred from these aboveground measurements.

5. Calculating the average change in carbon stocks for the entire project area is easiest if each sampling site encompasses the same area.

6. Subplots do not need to be the same shape—they can be square as well as round, for example—but their size should reflect the type of biomass. For example, if a project expects to grow few trees over 30 centimeters in diameter, quantifiers may want to install larger subplots for trees 15 to 30 centimeters in diameter because this size range will encompass most of the carbon sequestration and is thus worth measuring fairly precisely. Some subplots designated for measuring a given carbon pool might include no objects of that type, while others might contain more than a dozen. Such variation significantly affects the precision of the quantification system, but it may be unavoidable if the forest is lightly managed or if the types of biomass on a plot change over time.

7. Litter is undecomposed organic material lying on the ground, having piece sizes that are smaller than the minimum size of woody debris. Typically litter is leaves, twigs,

and bark. Duff is partially composed organic material lying on the ground surface. Typically the individual pieces of organic material in duff are visually distinguishable but matted together by fungal strands.

8. Note that this method is for estimating the mass of litter and duff which resides in the soil O horizon. Methods for measuring carbon in this layer are discussed in Chapter 7.

9. The density used in biomass equations is often reported as a *specific gravity* rather than a density. The specific gravity of a substance is simply expressed in units such that the density is given as a proportion of density divided by the density of the material relative to water. Because the density of water is 1 gram per cm³, the specific gravity of a substance is just a shorthand way of expressing the density of that substance in grams per cm³ without having to write down the units.

10. The Walkley-Black wet digestion method was formerly used to measure organic carbon in soil, but that approach is not sufficiently accurate. Also, it can underestimate carbon content and produces toxic waste.

11. Jenkins (2005) provides a comprehensive database of equations that predict the biomass of North American trees as a function of diameter. Equations that use both height and diameter, however, are strongly recommended. Biomass equations that use only diameter and species should be used only if no appropriate equation that uses both height and diameter is available and the project can not afford to develop the needed equations.

12. Aldred and Alemdag (1988) provide guidance on determining whether an existing tree biomass equation is appropriate.

13. See Appendix 1.

Chapter 7: Carbon Sequestered in Soil

1. This is equivalent to about 100,000 tons of CO_2 (see Chapter 2).

2. Paired sampling entails measuring the carbon stock on each plot at an earlier time and a later time, finding the change on each plot, and doing statistical analysis of the set of observed changes.

3. 1 hectare = 2.47 acres.

4. By chance, a plot center could sit very close to the boundary of the project land. Strict sampling theory would *fold* the plot along the property boundary and fold the part of the plot outside the boundary back within it. An acceptable alternative is to place a buffer on the south and east portion of the plot. This buffer can be between 4 and 9 meters wide, depending on the number of cores collected and the spacing between the cores.

5. In most soils, core-to-core variability of carbon content does not seem to rise significantly as core diameter and the volume of sampled soil shrink. Smaller-diameter corers are easier to insert into the ground, but the smaller the diameter is, the smaller the rock size is that can prevent crews from extracting intact cores. Smaller corers may have a greater tendency to compact soil, but they simplify transport and processing of soil. Rectangular corers are not recommended because the samples they extract seem to vary under many conditions.

6. Lichtenstein 1982; McClelland 1994.

7. Using subjective estimates of variability instead of empirical data for Monte Carlo modeling is little better than relying on expert opinion. For an example of a Monte Carlo analysis for land-use and management impacts on U.S. agricultural soils, see Ogle 2003.

8. The Walkley-Black wet digestion method was formerly used to measure organic carbon in soil, but that approach underestimates carbon content and produces toxic waste.

9. See Appendix 1 on sampling for a discussion of stratification.

10. See Appendices 3 and 4 for more on statistics and inadvertent emissions, and Chapters 5 and 10 for more on baselines and leakage.

Chapter 8: Greenhouse Gas Emissions from Manure

1. Methods for estimating the net impact of using biomass for fuel can be complex, however, and developers need to contract with the owner of the displaced fossil-fuel–fired facility to establish ownership of any resulting offsets.

2. Project emissions are the amount of gas emitted, not the amount produced.

3. Even management practices that produce high nitrous oxide emissions, such as pasture spreading, have a small overall warming impact per unit of manure.

4. A system that captures and burns methane may not reduce greenhouse emissions if it replaces a dry manure-management system because the amount of methane leaked from a digester system may be greater than unmanaged emissions from the dry management system.

5. Quantifiers should check the literature at the time of the analysis for the most up-to-date values.

6. See IPPC (1996), Table 4-8, p. 4.25.

7. For an equation for estimating emissions as a function of temperature, see U.S. EPA (2004). When making such calculations, quantifiers should use the temperature of the manure slurry, not the ambient air temperature, as the EPA does.

8. See IPCC (2000) and EPA (2003) for information on how to estimate emissions from feed inputs.

9. This number represents total dry matter, including ash content. In contrast to the IPCC method, manure fractions are not calculated or measured.

10. The 100 GWP of CH_4 assumes that the CH^4 will persist in the atmosphere for a relatively short time (8–12 years, on average) and then decompose to CO_2. The CO_2 then remains in the atmosphere for the rest of the 100-year accounting period. The GWP of CH_4 already counts the C in the CH_4 as being CO_2 in the atmosphere for about 90 percent of the accounting period. Burning CH_4 by the project changes the proportion of the 100-year accounting period from about 90 percent of the accounting period to 100 percent of the accounting period. This difference is subsumed in the uncertainty of the actual value of the 100-year GWP, so no further deduction is warranted for the extra few years that the C is in CO_2 in the atmosphere during the 100-year accounting period.

11. See IPCC (1996), which estimates the proportion of potential methane production that is lost, assuming that the digester converts manure to methane efficiently.

12. See IPCC (1996) and (2000), and U.S. EPA (2003) for similar methods. Documents from the Kyoto Protocol's Clean Development Mechanism (CDM) show how approved manure projects made these calculations. See http://cdm.unfccc.int.

13. The factors used to predict the rate of methane or nitrous oxide production as a function manure type and management system are highly uncertain. Agencies could develop better factors by establishing benchmark sites for measuring emissions from different manure management systems in different climates. Such efforts—while requiring substantial work—would help manure projects quantify their emissions more reliably.

14. The 2 percent rate is from IPCC (2000). The rate assumes that 1.25 percent of nitrogen in fertilizer is emitted from fields, 30 percent is leached, and 2.5 percent of the leached nitrogen converts to nitrous oxide in rivers. Adding 1.25 percent direct emissions and 0.75 percent emissions from streams yields 2 percent of applied nitrogen emitted as nitrous oxide.

Chapter 10: Estimating Leakage or Off-Site Emissions

1. If a facility operating under a regulated cap displaces emissions to other facilities operating under the cap, then those facilities must account for the displaced emissions, and thus the emissions are not considered leakage.

2. Economists often distinguish between two types of displacements: market leakage and activity shifting. This chapter focuses primarily on market leakage. See Appendix 20 for a brief discussion of these two types of displacements.

3. Leakage that increases offsite GHG emissions or reduces carbon stocks is known as *negative leakage*, and is subtracted from a project's net greenhouse benefit. Projects can also create *positive leakage* if they spur offsite cuts in GHG emissions. For example, if a project entails switching from plowing to no-till cropping, other farmers—observing that production costs are lower and yields in dry years are greater on project lands—might also switch to no-till. Because the resulting GHG benefits occur outside the project boundary, however, project developers do not own them, and thus cannot count them in calculating offsets. If they could be counted, it would raise the possibility of a given offset being counted twice. Of course, project developers could sign a contract with other landowners to bring their activities into the project, in which case they could produce offsets (see Appendix 21).

4. Project activities can result in spill-over emissions beyond project boundaries. This occurs when nitrogen fertilizer is applied to project lands, leaches into streams, and produces downstream nitrous oxide emissions. This also occurs with fugitive emissions, such as methane leaks from a facility designed to capture methane from decaying manure. To avoid confusion, these types of processes are not treated as leakage, but are accounted for in emissions from the project itself.

5. For a discussion of projects that increase demand, see Appendix 21.

6. Because demand decreases as price increases, E—the price elasticity of demand—is a negative number.

7. The equation in 10.3 is based on the simplest (comparative statics) method using elasticity estimates. More comprehensive economic modeling can be employed that will likely provide more precision. Murray et al. (2004) discuss these methods.

8. In Equation 2.2, leakage is expressed as a fraction or proportion of the total NET GHG Benefit.

9. *The magnitude of E* simply means expressing E as a positive number instead of a negative number. In our example, $E = -0.06$, so the magnitude of E is $+0.06$.

10. The literature on methods for estimating elasticities is extensive. For information on using a multiple-parameter regression to calculate supply and demand elasticities, see Marquez (2002), Greene (2000), and Edgerton (1996).

11. Recall that if $C_{proj} = C_{out}$, these parameters drop out of Equation 10.3 and thus are not needed.

12. The marginal rate of production is the amount of

production per unit of area on the next unit of land brought in to or out of production. The marginal rate of production is often—but not always—less than the average rate of production because more productive lands are usually already in production.

13. If marginal rates of production are employed, then C_{out} (above) should be estimated using data from marginal lands—and not average lands—outside the project.

Chapter 11: Verifying and Registering Offsets

1. These auditing principles are shared with the field of financial accounting. For a review of such principles, see also *Professional Standards*, issued annually by the American Institute of Certified Public Accountants, which includes statements on auditing practices from the Financial Accounting Standards Board. Several textbooks also provide introductions to auditing practices, including *Auditing and Assurance Services: An Integrated Approach* (Arins 2002).

2. Regulators issue emissions allowances, or caps, for each accounting period. Offsets create new opportunities for regulated emitters to meet their caps.

Appendix 1: Developing a Sampling Strategy

1. The standard deviation of the mean is defined as the square root of the variance in the measurements of the mean. The standard error of the estimate of the mean is the standard deviation divided by the square root of the number of sampling sites. Most spreadsheet programs allow users to automatically calculate the mean, standard deviation, and coefficient of variation of any dataset. For more information, see Appendix 3.

Appendix 2: Quantifying Inadvertent Emissions

1. Rates are calculated from information on emissions in the U.S. Inventory of Greenhouse Gases (EPA 2006).

2. Combustion efficiency is the proportion of fuel that is burned in a fire. If the amount of biomass has been measured before a fire, combustion efficiency can be calculated by measuring the amount of biomass remaining after the fire, and finding the proportion of the original biomass that has disappeared.

Appendix 5: Categorical Additionality and Barrier Tests

1. Under the CDM process, the Executive Board has approved project methodologies and consolidated methodologies for some sectors, and the Methodology Panel has published guidance on establishing guidelines. Approved project methodologies may be used as a template for quantifying offsets from new projects. Official Kyoto Protocol documents, including CDM documents, are available on the web. At this writing, the Executive Board and Methodologies Panel are only beginning to consider methods for determining baselines for projects that mitigate GHG emissions by changing land-use practices, and have not approved any methodologies for counting the benefits of these practices.

2. Regulators can limit the impact of different assumptions on additionality by requiring projects to use the same assumptions to establish the baseline.

3. For an analysis of the difficulties of documenting barriers to a forestry project under the Clean Development Mechanism, see Ellis (2003).

4. Barriers based on regulatory prohibitions are an exception, because they can usually be objectively demonstrated.

Appendix 6: Using Periodic Transition Rates

1. Because *FRL* is raised exponentially to determine the fractional coverage of a land-management practice, it should be carried to at least five significant digits.

Appendix 16: When Soil Density Changes

1. 1 hectare = 2.47 acres.

2. As Chapter 7 notes, rocks are checked for carbon. If carbon is found, rock fragments in soil samples are ground and added to the fine soil sample before sample mass and carbon content are measured.

Appendix 17: Determining Mass-Specific Ratios

1. The methodology recommended here for manure is not as rigorous as that recommended for woody material in Chapter 7. Because of the large variability in the dry matter content of manure slurry, the more precise methods recommended for woody matter would not be cost-effective.

Appendix 18: Calculating Emissions from Manure

1. The most recent IPCC global warming potential for methane is 23, but CDM projects implemented under the Kyoto Protocol use the older GWP of 21. Quantifiers should check the literature or the rules of a regulatory or market system for the most up-to-date or recommended GWP.

2. The most recent IPCC global warming potential for nitrous oxide is 296, but CDM projects implemented under the Kyoto Protocol use the older GWP of 310. Quantifiers should check the literature or the rules of a regulatory or market system for the most up-to-date or recommended GWP.

Appendix 27: Sample Field Protocol

1. *Taring* is setting a scale so it reads zero when holding an empty container that will be used to hold material being weighed. Be sure to re-tare the scale when changing containers.

2. For projects in the United States, it is recommended that plant species codes match codes used in the USDA Plants Database, http://plants.usda.gov/.

Bibliography

Adams, Darius M., et al. 1996. The forest and agricultural sector optimization model (FASOM): model structure and policy applications Res. Pap. PNW-RP-495. Portland, Ore.: U.S. Department of Agriculture, Forest Service, Pacific Northwest Research Station. http://agecon2/tamu.edu/people/faculty/mccarl-bruce/papers/503.pdf (retrieved 15 February 2007).

———. 2005. FASOMGHG Conceptual Structure and Specification: 2005 Documentation. http://agecon2/tamu.edu/people/faculty/mccarl-bruce/papers/1212FASOMGHG_doc.pdf (retrieved 15 February 2007).

Adams, Darius M., and R. W. Haynes. 1996. The 1993 timber assessment market model: Structure, projections, and policy simulations. General Technical Report PNW-GTR-368. Portland, Ore.: U.S. Department of Agriculture, Forest Service, Pacific Northwest Research Station.

Adams, M. B., and D. R. Owens. 2001. Specific gravity of coarse woody debris for some central Appalachian hardwood forest species. *Research Paper NE-716* Newtown Square, Penn.: U.S. Department of Agriculture, Forest Service, Northeastern Research Station.

Albaugh, T. J., H. L. Allen, P. M. Dourgherty, and K. H. Johnsen. 2004. Long-term growth responses of loblolly pine to optimal nutrient and water resource availability. *Forest Ecology and Management* 192:3–19.

Aldred, A. H., and I. S. Alemdag. 1988. Guidelines for Forest Biomass Inventory. *Information Report PI-X-77.* Canadian Forestry Service, Petawawa National Forestry Institute.

American Institute of Certified Public Accountants (AICPA). 2003. *AICPA Professional Standards.* New York: AICPA.

Arins, A. A., M. S. Beasley, and R. J. Elder. 2002. *Auditing and Assurance Services: An Integrated Approach.* 9th edition. New York: Prentice-Hall.

Attfield, C. L. F. 1985. Homogeneity and endogeneity in systems of demand equations. *Journal of Econometrics* 27:197–209.

Avery, T. E., and H. E. Burkhart. 1994. *Forest Measurements.* 4th edition. New York: McGraw-Hill.

Baldwin, V., et al. 2000. The effects of spacing and thinning on stand and tree characteristics of 38-year-old Loblolly Pine. *Forest Ecology and Management* 137:91–102.

Birdsey, R. A. 1992. Carbon storage and accumulation in United States forest ecosystems. General Technical Report WO-59. Washington: U.S. Department of Agriculture, Forest Service.

Birdsey, R. A. 1996. Carbon storage for major forest types and regions in the conterminous United States. *Forests and Global Change,* vol. 2, *Forest Management Opportunities for Mitigating Carbon Emissions*, ed. R. N. Sampson and D. Hair. Washington: American Forests.

Birdsey, R. A. 1996. Appendix 2–4: Regional estimates of timber volume and forest carbon. In *Forests and Global Change,* vol. 2, *Forest Management Opportunities for Mitigating Carbon Emissions*, ed. R. N. Sampson and D. Hair. Washington: American Forests.

Birdsey, R. A., and G. M. Lewis. 2003. Carbon in U.S. Forests and Wood Products, 1987–1997: State-by-State Estimates. General Technical Report NE-310. Newtown Square, Penn.: U.S. Department of Agriculture, Forest Service, Northeastern Research Station.

Bledsoe, C. S., et al. 1999. Measurement of static root parameters: Biomass, length, and distribution in the soil profile. *Standard Soil Methods for Long-Term Ecologi-*

cal Research, ed. Robertson, G. Philip, D. C. Coleman, C. S. Bledsoe, and P. Sollins. New York: Oxford University Press.

Böhm, W. 1979. *Methods of Studying Root Systems.* Berlin: Springer.

Boscolo, M., J. R. Vincent, and T. Panayotou. 1998. Discounting Costs and Benefits in Carbon Sequestration Projects. *Discussion Paper 638.* Cambridge: Harvard University, Institute for International Development.

Brady, N. C., and R. R. Weil. 1996. *The Nature and Properties of Soils,* 11th edition. Upper Saddle River, N.J.: Prentice-Hall.

Brooks, H. G., S. V. Aradhyula, and S. R. Johnson. 1992. Land quality and producer participation in U.S. commodity programs. *Review of Agricultural Economics* 14, no. 1:105–15.

Brown, J. K. 1974. Handbook for Inventorying Downed Woody Material. General Technical Report INT-16. Ogden, Utah: U.S. Department of Agriculture, Forest Service, Intermountain Forest and Range Experiment Station.

———. 1976. Estimating shrub biomass from basal stem diameters. *Canadian Journal of Forest Research* 6:153–58.

Brown, J. K., and P. J. Roussopoulos. 1974. Eliminating biases in the planar intercept method for estimating volumes of small fuels. *Forest Science* 20:350–56.

Burtraw, D., et al. 2001. *The Effect of Allowance Allocation on the Cost of Carbon Emission Trading.* Washington: Resources for the Future.

Campbell, C. A., et al. 1999. Soil quality: Effect of tillage and fallow frequency in a silt loam in southwestern Saskatchewan. *Soil Biology and Biochemistry* 32, no. 1:1–7.

Canary, J. D., R. B. Harrison, R. E. Edmonds, and H. N. Chappell. 2000. Carbon sequestration following repeated urea fertilization of second-growth Douglas-fir stands in western Washington. *Forest Ecology and Management* 138:225–32.

CDM: Executive Board. 2005. Tool for the demonstration and assessment of additionality (version 02). http://cdm.unfccc.int/methodologies/PAmethodologies/AdditionalityTools/Additionality_tool.pdf (retrieved 15 February 2007).

Cicerone, R. J., C. C. Delwiche, and J. D. Shetter. 1983. Seasonal variation of methane flux from a California rice paddy. *Journal of Geophysical Research.* 88:11022–24.

Cicerone, R. J., C. C. Delwiche, T. C. Tyler, and P. R. Zimmermann. 1992. Methane emission from Californian rice paddies with varied treatment. *Global Biogeochemical Cycles* 6:233–48.

Cihacek, L. J. 2003. Monitoring carbon sequestered by soil. *Meeting of the Emissions Marketing Association,* Miami, 21–23 September. www.emissions.org/conferences/fallconference03/cd.php (retrieved 15 February 2007).

Clark, A. 1987. Summary of biomass equations available for softwood and hardwood species in the southern United States. *Estimating Tree Biomass Regressions and Their Error: Proceedings of the Workshop on Tree Biomass Regression Functions and their Contribution to the Error of Forest Inventory Estimates.* General Technical Report NE-GTR-117, ed. E. H. Wharton and T. Cunia. Broomall, Penn.: U.S. Department of Agriculture, Forest Service, Northeastern Forest Experiment Station.

Clark, A., D. R. Phillips, and D. J. Frederick. 1986. Weight, Volume, and Physical Properties of Major Hardwood Species in the Piedmont. *Research Paper SE-255.* Asheville, N.C.: U.S. Department of Agriculture, Forest Service, Southeastern Forest Experiment Station.

Cochran, P. H., J. W. Jennings, and C. T. Youngberg. 1984. *Biomass Estimators for Thinned Second-Growth Ponderosa Pine Trees (Research Note PNW-415)* Portland, Ore.: U.S. Department of Agriculture, Forest Service, Pacific Northwest Forest and Range Experiment Station.

Cole, C. V., K. Flach, J. Lee, D. Sauerbeck, and B. Stewart. 1993. Agricultural sources and sinks of carbon. *Water, Soil, and Air Pollution* 70:111–22.

Conant, R. T., G. R. Smith, and K. Paustian. 2003. Spatial variability of soil carbon in forested and cultivated sites: Implications for change detection. *Journal of Environmental Quality* 32, no. 1:278–86.

Congressional Budget Office. 2000. *Who Gains and Who Pays under Carbon-Allowance Trading? The Distributional Effects of Alternative Policy Designs.* Washington: Congressional Budget Office.

Conservation Technology Information Center. 2004. *2002 National Crop Residue Management Survey.* www.ctic.purdue.edu/CTIC/CTIC.html (retrieved 2 March 2007).

De Jonge, L., G. Mulder, and S. Greiner. 2004. Proposal on CDM additionality tests, Netherlands Ministry of Environment (VROM), World Bank Carbon Finance Business, and Carboncredits.nl team of SenterNovem. www2.vrom.nl/docs/internationaal/proposal_on_CDM%20Additionally%20Tests.pdf (retrieved 2 February 2007).

Den Elzen, M. G. J., and M. Meinshausen. 2006. Multi-Gas Emission Pathways for Meeting EU 2°C Climate Target. *Avoiding Dangerous Climate Change.* Cambridge: Cambridge University Press.

Denier van der Gon, H., et al. 1993. Controlling factors of methane emission from rice fields. *Proceedings of WISE Workshop, August 24–27, 1992.* Wageningen, Netherlands.

———. 1996. Release of entrapped methane from wetland rice fields upon soil drying. *Global Biogeochemical Cycles* 10:1–7.

Dick, W.A., and J. T. Durkalski. 1997. No-tillage production agriculture and carbon sequestration in a typic Fragiudalf soil of northeastern Ohio. *Management of Carbon Sequestration in Soil*, ed. R. Lal , J. M. Kimble, R. F. Follett, and B. A. Stewart. Boca Raton: CRC.

Donald, P. R. B., and J. Martel. 1997. Impact of tillage practices on organic carbon and nitrogen storage in cool, humid soils of eastern Canada. *Soil Tillage Research* 41:191–201.

Edgerton, D. L., et al. 1996. *The Econometrics of Demand Systems with Applications to Food Demand in the Nordic Countries.* Doordrecht, Netherlands: Kluwer.

Ellis, Jane. 2003. Forest projects: lessons learned and implications for CDM modalities. Organisation for Economic Co-operation and Development and International Energy Agency. www.oecd.org/dataoecd/24/15/2956438.pdf (retrieved 2 February 2007).

Galle, B., et al. 2000. Measurements of ammonia emissions from spreading of manure using gradient FTIR techniques. *Atmospheric Environment* 34:4907–15.

Gibbard, S., et al. 2005. Climate Effects of Global Land Cover Change. *Geophys. Res. Lett.* 32: L23705, doi: 10.1029/2005GLO24550.

Gifford, Roger M., and Michael L. Roderick. 2003. Soil carbon stocks and bulk density: Spatial or cumulative mass coordinates as a basis of expression? *Global Change Biology* 9:1507–14.

Greene, W. H. 2000. *Econometric Analysis,* 4th edition. London: Prentice-Hall.

Gregoire, T. G., H. T. Valentine, and G. M. Furnival. 1995. Sampling methods to estimate foliage and other characteristics of individual trees. *Ecology* 76, no. 4:1181–94.

Grier, C.C., et al. 1981. Biomass distribution and above- and below-ground production in young and mature *Abies amabilis* zone ecosystems of the Washington Cascades. *Canadian Journal of Forest Research* 11:155–67.

Hamburg, S. P., et al. 1997. Estimating the carbon content of Russian forests; a comparison of phytomass/volume and allometric projections. *Mitigation and Adaptation Strategies for Global Change* 2:247–65.

Harmon, M. E., S. L. Garman, and W. K. Ferrell. 1996. Modeling historical patterns of tree utilization in the Pacific Northwest: carbon sequestration implications. *Ecological Applications* 6, no. 2:641–52.

Harmon, M. E., J. M. Harmon, W. K. Ferrell, and D. Brooks. 1996. Modeling carbon stores in Oregon and Washington Forest Products, 1900–1992. *Climatic Change* 33:521–50.

Harmon, M. E. and K. Lajtha. 1999. Analysis of detritus and organic horizons for mineral and organic constituents. *Standard Soil Methods for Long-Term Ecological Research*, ed. G. P. Robertson, D. C. Coleman, C. S. Bledsoe, and P. Sollins. New York: Oxford University Press.

Harmon, M. E., and B. Marks. 2002. Effects of silvicultural practices on carbon stores in Douglas-fir/western hemlock forests in the Pacific Northwest, U.S.A.: Results from a simulation model. *Canadian Journal of Forest Research* 32:863–77.

Harmon, M. E., and J. Sexton. 1996. Guidelines for Measurements of Wood Detritus in Forest Ecosystems. *U.S. LTER Publication No. 20.* Seattle: University of Washington, College of Forest Resources.

Harrison, R. B., et al. 2003. Quantifying deep-soil and coarse-soil fractions: Avoiding sampling bias. *Soil Science Society of America Journal* 67:1602–6.

Harrison, R. M., et al. 1995. Effect of fertilizer application on NO and N_2O fluxes from agricultural fields. *Journal of Geophysical Research* 100:25923–31.

Harvey, A. C. 1990. *The Econometric Analysis of Time Series*, 2nd edition Cambridge: MIT Press.

Haynes, R. W. 2003. An Analysis of the Timber Situation in the United States, 1952 to 2050. General Technical Report PNW-GTR-560. Portland, Ore.: U.S. Department of Agriculture, Forest Service, Pacific Northwest Research Station.

Hoag, D. L., B. A. Babcock, and W. E. Foster. 1993. Field-level measurement of land productivity and program slippage. *American Journal of Agricultural Economics* 75:181–89.

Houghton, R. A. 1999. The Annual Net Flux of Carbon to the Atmosphere from Changes in Land Use, 1850–1990, *Tellus* 51B.

House, J. I., I. C. Prentice, and C. Le Quéré. 2002. Maximum Impacts of Future Reforestation or Deforestation on Atmospheric CO_2. *Global Change Biology* 8, no. 11:1047–52.

Howard, J. L. 2004. U.S. forest products annual market review and prospects, 2001–2004. *Research Note FPL-RN-0292.* Madison, Wis.: U.S. Department of Agriculture, Forest Service, Forest Products Laboratory.

Intergovernmental Panel on Climate Change (IPCC). 1995. *Climate Change 1995: The Science of Climate Change.* Cambridge: Cambridge University Press.

———. 1996. *Greenhouse Gas Inventory Reporting Instructions: IPCC Guidelines for National Greenhouse Gas Inventories, Revised,* vol. 3: *Reference Manual.* Kanagawa, Japan: IPCC National Greenhouse Gas Inventories Programme Technical Support Unit. www.ipcc-nggip.iges.or.jp/public/2006gl/index.htm (retrieved 11 May 2007).

———. 2000. *Good Practice Guidance and Uncertainty Management in National Greenhouse Gas Inventories,* ed. J. Penman, D. Kruger, I. Galbally, T. Hiraishi, B. Nyenzi, S. Emmanul, L. Buendia., R. Hoppaus, T. Martinsen, J. Meijer, K. Miwa, and K. Tanabe. Japan: Institute for Global Environmental Strategies. www.ipcc-nggip.iges.or.jp/public/gp/gpgaum.htm (retrieved 2 February 2007).

———. 2001. *Climate Change 2001: The Scientific Basis.* Cambridge: Cambridge University Press.

———. 2003. *Good Practice Guidance for Land Use, Land-Use Change and Forestry.* Kanagawa, Japan: Intergovernmental Panel on Climate Change, National Greenhouse Gas Inventories Programme Technical Support Unit. www.ipcc-nggip.iges.or.jp/public/gpglulucf/gpglulucf.htm (retrieved 11 May 2007).

Izaurralde, R. C., et al. 1998. Scientific challenges in developing a plan to predict and verify carbon storage in Canadian prairie soils. *Management of Carbon Sequestration in Soil,* ed. R. Lal, J. M. Kimble, R. F. Follett, and B. A. Stewart. Boca Raton: CRC.

Jenkins, J. C., et al. 2003. National-scale biomass estimators for United States tree species. *Forest Science* 49, no. 1:12–35.

———. 2005. Comprehensive Database of Diameter-based Biomass Regressions for North American Tree Species. General Technical Report NE-319. Newtown Square, Penn.: U.S. Department of Agriculture, Forest Service, Northeastern Research Station.

Jenkins, J. C., R. A. Birdsey, and Y. Pan. 2001. Biomass and NPP estimation for the mid-Atlantic region (USA) using plot-level forest inventory data. *Ecological Applications* 11:1174–93.

Johnson, E. W. 2000. *Forest Sampling Desk Reference.* Boca Raton: CRC.

Keller, F., et al. 2006. Methane emissions form terrestrial plants under aerobic conditions. *Nature* 439:187–91.

Kim, M. K. 2004. *Economic Investigation of Discount Factors for Agricultural Greenhouse Gas Emissions Offsets.* Ph.D. dissertation, Department of Agricultural Economics, Texas A&M University.

Kimmins, J. P. 1987. *Forest Ecology.* New York: Macmillan.

Koch, Peter. 1989. Estimates by species group and region in the USA of: I. Below-ground root weight as a percentage of ovendry complete-tree weight; and II. Carbon content of tree portions. Corvallis, Mont.: Wood Science Laboratory. [Manuscript on file with U.S. Department of Agriculture, Forest Service, Northeastern Research Station, Newtown Square, Penn.]

Krankina, O. N., and M. E. Harmon. 1994. The impact of intensive forest management on carbon stores in forest ecosystems. *World Resources Review* 6, no. 2:161–77.

Lashof, D. 2001. *Rising Emissions: The Failure of Voluntary Commitments and Reporting to Reduce U.S. Electric Industry CO_2 Emissions.* New York: Natural Resources Defense Council.

Lewandrowski, J., et al. 2004. Economics of Sequestering Carbon in the U.S. Agricultural Sector. Technical Bulletin No. 1909. Washington: U.S. Department of Agriculture, Economic Research Service.

Li, C. 2000. Modeling trace gas emissions from agricultural ecosystems. *Nutrient Cycling in Agroecosystems* 58:259–76.

———. 2001. Biogeochemical concepts and methodologies: Development of the DNDC model. *Quaternary Sciences* 21:89–99.

Li, C., et al. 2004. Modeling Greenhouse Gas Emissions from Rice-Based Production Systems: Sensitivity and Upscaling. *Global Biogeochemical Cycles.* 18:GB1043 [doi:10.1019/2003GB002045].

Li, C., J. Aber, F. Stange, K. Butterbach-Bahl, and H. Papen. 2000. A process-oriented model of N_2O and NO emissions from forest soils: 1. Model development. *Journal of Geophysical Research* 105, no. 4:4369–84.

Li, C., S. Frolking, and T. A. Frolking. 1992. A model of nitrous oxide evolution from soil driven by rainfall events: 1. Model structure and sensitivity. *Journal of Geophysical Research* 97: 9759–76.

Li, C., V. Narayanan, and R. Harriss. 1996. Model estimates of nitrous oxide emissions from agricultural lands in the United States. *Global Biogeochemical Cycles* 10:297–306.

Lichtenstein, S., B. Frischhoff, and L. D. Phillips. 1982. Calibration of probabilities: The state of the art to 1980. *Judgment under Uncertainty: Heuristics and Biases,* ed. D. Kahneman, P. Slovic, and A. Tversky, New York: Cambridge University Press.

Marquez, J. 2002. *Estimating Trade Elasticities*. Dordrecht, Netherlands: Kluwer.

McAlister, R. H., H. R. Powers, and W. D. Pepper. 2000. Mechanical properties of stemwood and limbwood of seed orchard Loblolly pine. *Forest Products Journal* 50, no. 10:91–94.

McCarty, G. W., N. N. Lyssenko, and J. L. Starr. 1998. Short-term changes in soil carbon and nitrogen pools during tillage management transition. *Soil Science Society of America Journal* 62:1564–71.

McClellend, A. G. R., and F. Bolger. 1994. The calibration of subjective probability: Theories and models, 1980–1994. *Subjective Probability*, ed. G. Wright and P. Ayton. Chichester: John Wiley and Sons.

Means, J. E., et al. 1994. Software for Computing Plant Biomass: BIOPAK Users Guide. General Technical Report PNW-GTR-340. Portland, Ore.: U.S. Department of Agriculture, Forest Service, Pacific Northwest Research Station.

Means, J. E., P. C. MacMillan, and K. Cromack Jr. 1992. Biomass and nutrient content of Douglas-fir logs and other detrital pools in an old-growth forest, Oregon, U.S.A. *Canadian Journal of Forest Research* 22:1536–46.

Meyer, W. H. 1938. Yield of even-aged stands of ponderosa pine. Technical Bulletin No. 630. Washington: U.S. Department of Agriculture.

Mosier, A. R., et al. 1996. Nitrous oxide emissions from agricultural fields: Assessment, measurement and mitigation. *Plant and Soil* 181:95–108.

———. 1998. Mitigating agricultural emissions of methane. *Climatic Change* 40:39–80.

———. 1998. Mitigating agricultural emissions of nitrous oxide. *Climatic Change* 40:7–38.

Murray, B. C., B. A. McCarl, and H. C. Lee. 2004. Estimating leakage from forest carbon sequestration programs. *Land Economics* 80, no. 1:109–24.

Nadelhoffer, K. J., J. D. Aber, and J. M. Melillo. 1985. Fine roots, net primary production, and soil nitrogen availability: a new hypothesis. *Ecology* 66:1377–90.

Naidu, S., E. H. DeLucia, and R. B. Thomas. 1998. Contrasting patterns of biomass allocation in dominant and suppressed loblolly pine. *Canadian Journal of Forest Research* 28:1116–24.

Nelson, D. W., and L. E. Sommers. 1982. Total carbon, organic carbon, and organic matter. *Methods of Soil Analysis: Chemical and Microbiological Properties,* Monograph 9, ed. A. L. Page, R. H. Miller, and D. R. Kenney. Madison, Wis.: American Society of Agronomy.

Neue, H. U., R. S. Latin, R. Wassmann, J. B. Aduna, C. R. Alberto, and M. J. F. Andales. 1994. Methane emission from rice soils of the Philippines. *CH4 and N2O: Global Emissions and Controls from Rice Fields and Other Agricultural and Industrial Sources,* ed. K. Minami, A. Mosier, and R. Sass. Tokyo: Yokendo.

Newbold, R. A., V. C. Baldwin Jr., and G. Hill. 2001. Weight and Volume Determination for Planted Loblolly Pine in North Louisiana. Research Paper SRS-26. Asheville, NC: U.S. Department of Agriculture, Forest Service, Southern Research Station.

Ogle, S. M., F. Jay Breidt, M. D. Eve, and K. Paustian. 2003. Uncertainty in estimating land use and management impacts on soil organic carbon storage for U.S. agricultural lands between 1982 and 1997. *Global Change Biology* 9: 1521–42.

Olander, L. 2006. *Do Recent Scientific Findings Undermine the Climate Sequestration Benefits of Forests?* Durham, N.C.: Nicholas Institute. www.nicholas.duke.edu/institute/methanewater.pdf (retrieved 7 March 2007).

O'Neill, B. C., and M. Oppenheimer. 2002. Climate Change: Dangerous Climate Impacts and the Kyoto Protocol. *Science* 296:1971–72.

Ovington, J. D. 1956. The composition of tree leaves. *Forestry* 29:22–29.

Palmer, C. J., W. D. Smith, and B. L. Conkling. 2002. Development of a protocol for monitoring status and trends in forest soil carbon at a national level. *Environmental Pollution* 116:S209–S219.

Parker, R. C., and T. G. Matney. 1999. Comparison of optical dendrometers for prediction of standing tree volume. *Southern Journal of Applied Forestry* 23, no. 2:100–107.

Parresol, B. R. 1999. Assessing tree and stand biomass: A review with examples and critical comparisons. *Forest Science* 45, no. 4:573–93.

———. 2001. Additivity of nonlinear biomass equations. *Canadian Journal of Forest Research* 31:865–78.

Parton, W. J., D. S. Schimel, C. V. Cole, and D. S. Ojima. 1987. Analysis of factors controlling soil organic matter levels in Great Plains grasslands. *Soil Science Society of America Journal* 51:1173–79.

Paul, E. A., and F. E. Clark. 1989. *Soil Microbiology and Biochemistry*, 2nd edition. San Diego: Academic.

Peoples, M. B., A. R. Mosier, and J. R. Freney. 1995. Minimizing gaseous losses of nitrogen. *Nitrogen Fertilization in the Environment*, ed. P. Bacon. New York: Marcel Dekker.

Pillsbury, N. H., and M. L. Kirkley. 1984. Equations for Total, Wood, and Saw-Log Volume from Thirteen California Hardwoods. Research Note PNW-414. Portland,

Ore.: U.S. Department of Agriculture, Forest Service, Pacific Northwest Forest and Range Experiment Station.

Richards, K. R. 1997. The time value of carbon in bottom-up studies. *Critical Review in Environmental Science and Technology* 27 (special issue):S279–S292.

Row, C. 1996. Effects of selected forest management options on carbon storage. *Forests and Global Change,* vol. 2: *Forest Management Opportunities for Mitigating Carbon Emissions,* ed. R.N. Sampson and D. Hair. Washington: American Forests.

Row, C., and R. B. Phelps. 1996. Wood carbon flows and storage after timber harvest. *Forests and Global Change,* vol. 2: *Forest Management Opportunities for Mitigating Carbon Emissions,* ed. R. N. Sampson and D. Hair. Washington: American Forests.

Rygnestad, H., and R. Fraser. 1996. Land heterogeneity and the effectiveness of cap set-aside. *Journal of Agricultural Economics* 47, no. 2:255–60.

Santantonio, D., R. K. Hermann, and W. S. Overton. 1977. Rood biomass studies in forest ecosystems. *Pedobiologia* 17:1–31.

Schlesinger, W. H. 2000. Carbon sequestration in soils: some cautions amidst optimism. *Agriculture, Ecosystems and Environment* 82:121–27.

Seale, J., Jr., A. Regmi, and J. Bernstein. 2003. International Evidence on Food Consumption Patterns. *Technical Bulletin 1904* [October]. Washington: U.S. Department of Agriculture. www. ers.usda.gov/publications/tbl904/ (retrieved 2 February 2007).

Six, J., E. T. Elliott, and K. Paustian. 1999. Aggregate and soil organic matter dynamics under conventional and no-tillage systems. *Soil Science Society of America Journal* 63:1350–58.

Six, J., S. M. Ogle, F. J. Breidt, R. T. Conant, A. R. Mosier, and K. Paustian. 2004. The potential to mitigate global warming with no-tillage management is only realized when practised in the long term. *Global Change Biology* 10:155–60.

Skog, K. E., and G. A. Nicholson. 1998. Carbon cycling through wood products: The roe of wood and paper products in carbon sequestration. *Forest Products Journal* 48, nos. 7–8:75–83.

Smith, G. R. 2002. Designing Sampling Systems to Detect Carbon Stored in Soil or Forest Sequestration Projects. *U.S. Department of Agriculture Symposium on Natural Resource Management to Offset Greenhouse Gas Emissions,* Raleigh, N.C., 19–21 November 2002.

Smith, J. E., and L. S. Heath. 2002. A Model of Forest Floor Carbon Mass for United States Forest Types. Research Paper NE-722. Newtown Square, Penn.: U.S.

Department of Agriculture, Forest Service, Northeastern Research Station.

Smith, J. E., L. S. Heath, and J. C. Jenkins. 2003. Forest Volume-to-Biomass Models and Estimates of Mass for Live and Standing Dead Trees of U.S. Forests. General Technical Report NE-298. Newtown Square, Penn.: U.S. Department of Agriculture, Forest Service, Northeastern Research Station.

Smith, S. V., R. O. Sleezer, W. H. Renwick, and R. W. Buddemeier. 2005. Fates of eroded soil organic carbon: Mississippi Basin case study. *Ecological Applications* 15, no. 6:1929–40.

Smithwick, E. A. H, et al. 2002. Potential upper bounds of carbon stores in forests of the Pacific Northwest. *Ecological Applications* 12, no. 5:1303–17.

Soil Survey Laboratory. 1995. *Soil Survey Laboratory Information Manual.* Soil Survey Investigations Report no. 45, version 1.0. Lincoln, Neb.: U.S. Department of Agriculture, Natural Resources Conservation Service, National Soil Survey Center.

Sollins, P., et al. 1987. Patterns of log decay in old-growth Douglas-fir forests. *Canadian Journal of Forest Research* 17:1585–95.

Sollins, P., C. Glassman, E. A. Paul, C. Swanston, K. Lajtha, J. W. Heil, and E. T. Elliott. 1999. Soil carbon and nitrogen: pools and fractions. *Standard Soil Methods for Long-Term Ecological Research,* ed. G. P. Robertson, D. C. Coleman, C. S. Bledsoe, and P. Sollins. New York: Oxford University Press.

Stokey, E., and R. Zeckhauser. 1978. *A Primer for Policy Analysis.* New York: W. W. Norton.

Stumm, W., and J. J. Morgan. 1981. *Aquatic Chemistry: An Introduction Emphasizing Chemical Equilibria in Natural Waters,* 2nd edition. New York: John Wiley and Sons.

Taras, M. A. 1980. Aboveground Biomass of Choctawhatchee Sand Pine in Northwest Florida. Research Paper SE-210. Asheville, N.C.: U.S. Department of Agriculture, Forest Service, Southeastern Forest Experiment Station.

Thomas, C. E., et al. 1995. Biomass and taper for trees in thinned and unthinned longleaf pine plantations. *Southern Journal of Applied Forestry* 19, no. 1:29–35.

Trexler, M., and J. Shipley. 2004. Back to the Future. *Carbon Finance.* 10:12–13.

United Nations. 2002. Report of the Conference of the Parties on Its Seventh Session, Held at Marrakesh from 29 October to 10 November 2001. http://unfccc.int/resource/docs/cop7/13a02.pdf, 39 (retrieved 1 March 2007).

U.S. Department of Energy. 1994. *Sector-Specific Issues*

and Reporting Methodologies Supporting the General Guidelines for the Voluntary Reporting of Greenhouse Gases under Section 1605(b) of the Energy Policy Act of 1992. Washington: U.S. Department of Energy.

———. 2006. *Technical Guidelines for Voluntary Reporting of Greenhouse Gas Program,* part I: *Appendix: Forestry.* Washington: U.S. Department of Energy, Office of Policy and International Affairs.

U.S. Energy Information Administration. 2004. *Emissions of Greenhouse Gases in the United States, 2003.* DOE/EIA-0573. Washington: U.S. Department of Energy.

U.S. Environmental Protection Agency. 2001. *Inventory of U.S. Greenhouse Gas Emissions and Sinks, 1990–1999.* EPA 236-R-01-001 (6204N). Washington: U.S. Environmental Protection Agency, Office of Atmospheric Programs.

———. 2003. *Inventory of U.S. Greenhouse Gas Emissions and Sinks, 1990–2001.* Washington: U.S. Environmental Protection Agency, Office of Atmospheric Programs. yosemite.epa.gov/oar/globalwarming.nsf/content/ResourceCenterPublicationsGHGEmissions.html (retrieved 11 May 2007).

———. 2004. *Inventory of U.S. Greenhouse Gas Emissions and Sinks: 1990–2002.* EPA 430-R-03-004. Washington: U.S. Environmental Protection Agency, Office of Atmospheric Programs. yosemite.epa.gov/oar/globalwarming.nsf/content/ResourceCenterPublicationsGHGEmissions.html (retrieved 11 May 2007).

———. 2005. *Greenhouse Gas Mitigation Potential in U.S. Forestry and Agriculture.* EPA 430-R-05-006. Washington: U.S. Environmental Protection Agency, Office of Atmospheric Programs. www.epa.gov/sequestration (retrieved 2 February 2007).

———. 2006. *Inventory of U.S. Greenhouse Gas Emissions and Sinks, 1990–2004.* EPA 430-R-05-003. www.epa.gov/globalwarming/publications/emissions (retrieved 2 February 2007).

U.S. Senate. Senate Amendment No. 866 to H.R. 6, the Energy Policy Act of 2005, submitted as "Sense of the Senate on Climate Change." http://thomas.loc.gov/cgi-bin/query/F?r109:1:./temp/~r109j833vW:e119074: (retrieved 7 March 2007).

Valentine, H. T., L. M. Trotton, and G. M. Furnival. 1984. Subsampling trees for biomass, volume, or mineral content. *Forest Science* 30:673–81.

van Wagner, C. E. 1968. The line intercept method in forest fuel sampling. *Forest Science* 14, no. 1:20–26.

Vincent, K. R., and O. A. Chadwick. 1994. Synthesizing bulk density for soils with abundant rock fragments. *Soil Science Society of America Journal* 58:455–64.

Vitousek, P. M., and R. L. Sanford Jr. 1986. Nutrient cy-

cling in moist tropical forest. *Annual Review of Ecology and Systematics* 17:137–67.

Vogt, K. A., and H. Persson. 1991. Measuring growth and development of roots. *Techniques and Approaches in Forest Tree Ecophysiology,* ed. J. P. Lassoie and T. M. Hinckley. Boca Raton: CRC.

Vogt, K. A., D. J. Vogt, S. Brown, J. P. Tilley, R. L. Edmonds, W. L. Silver, and T. G. Siccama. 1995. Dynamics of forest floor and soil organic matter accumulation in boreal, temperate, and tropical forests. *Soil Management and Greenhouse Effect,* ed. R. Lal, J. Kimble, E. Levine, and B. A. Stewart. Boca Raton: CRC.

Ward, D. E., et al. 1996. Effect of fuel composition on combustion efficiency and emission factors for African savanna ecosystems. *Journal of Geophysical Research* 101:23569–76.

Waring, R. H., and J. F. Franklin. 1979. Evergreen coniferous forests of the Pacific Northwest. *Science* 204:1380–86.

Waring, R. H., and W. H. Schlesinger. 1985. *Forest Ecosystems: Concepts and Management.* Orlando: Academic.

Wassmann, R., H. Papen, and H. Rennenberg. 1993. Methane emission from rice paddies and possible mitigation strategies. *Chemosphere.* 26:201–17.

Wassmann, R., M. X. Wang, X. J. Shangguan, X. L. Xie, R. X. Shen, H. Papen, H. Rennenberg, and W. Seiler. 1993. First records of a field experiment on fertilizer effects on methane emission from rice fields in Hunan Province (PR China). *Geophysical Research Letters.* 20:2071–74.

Wear, D. N., and B. C. Murray. 2004. Federal timber restrictions, interregional spillovers, and the impact on US softwood markets. *Journal of Environmental Economics and Management* 47, no. 2:307–30.

Wenger, K. F. 1984. *Forestry Handbook,* 2nd edition. New York: John Wiley and Sons.

West, P. W. 2003. *Tree and Forest Measurement.* Berlin: Springer.

Wharton, E. H., and T. Cunia. 1987. Estimating Tree Biomass Regressions and Their Error: Proceedings of the Workshop on Tree Biomass Regression Functions and Their Contribution to the Error of Forest Inventory Estimates. General Technical Report NE-GTR-117. Broomall, Penn.: U.S. Department of Agriculture, Forest Service, Northeastern Forest Experiment Station.

World Resources Institute and the World Business Council for Sustainable Development. 2003. The Greenhouse Gas Protocol for Project Accounting. www.ghgprotocol.org/DocRoot/m1Tv5lnUuFTjYZx3x1ev/GHG_Project_Protocol.pdf (retrieved 2 March 2007).

Wu, J. J. 2000. Slippage effects of the Conservation Reserve Program. *American Journal of Agricultural Economics.* 82, no. 4:979–92.

Yanai, R. D., S. V. Stehman, M. A. Arthur, C. E. Prescott, A. J. Friedland, T. G. Siccama, and D. Binkley. 2003. Detecting change in forest floor carbon. *Soil Science Society of America Journal* 67:1583–93.

Yuancai, L., and B. R. Parresol. 2001. Remarks on height-diameter modeling. Research Note SRS-10. Asheville, N.C.: U.S. Department of Agriculture, Forest Service, Southern Research Station.

Zahn, J. A., J. L. Hatfield, Y. S. Do, A. A. DiSpirito, D. A. Laird, and R. L. Pfeiffer. 1997. Characterization of volatile organic emissions and wastes from a swine production facility. *Journal of Environmental Quality* 26:1687–96.

Zhou, Y., Z. Tao, and D. D. Du. 1994. Agricultural measure options for reducing methane emission from rice field. *Rural Eco-Environment.* 10:6–8.

Index